NEW DEVELOPMENTS IN MEDICAL RESEARCH

MEDICAL PHYSICS

NEW DEVELOPMENTS IN MEDICAL RESEARCH

Additional books in this series can be found on Nova's website
under the Series tab.

Additional e-books in this series can be found on Nova's website
under the e-book tab.

NEW DEVELOPMENTS IN MEDICAL RESEARCH

MEDICAL PHYSICS

MARCIN BALCERZYK
EDITOR

NOVA BIOMEDICAL

New York

For permission to use material from this book please contact us:
Telephone 631-231-7269; Fax 631-231-8175
Web Site: http://www.novapublishers.com

Library of Congress Cataloging-in-Publication Data

ISBN: 978-1-62257-590-9

Library of Congress Control Number: 2012943800

Published by Nova Science Publishers, Inc. † New York

Contents

Preface

Physics aims at explaining and describing as many processes of nature as possible. In this book we show several examples of its usefulness in medical environment. We start from basic processes like bacterial diffusion, eye vision, through preclinical applications, to radiation treatment planning. This sampling allows profound examination of powerful tools, which physics provide for medical field.

The book is intended for medical students at doctoral or postdoctoral level, students of medical physics of master level or upper, or for medical doctors who want to have a deeper insight into their specialty.

In Chapter I, Dr. Culea presents some of her research results in medical applications by using gas chromatography- mass spectrometry, especially the achievement of non-invasive or minimum invasive advanced methods of diagnosis and treatment of infants and children.

In Chapter II, Dr. Park focuses on the individual dynamics of a single microbe and review statistical mechanical approach for a system of such particles in relatively low concentration.

In Chapter III, Dr. Florin simulates the daylight perception by a normal eye and a Daltonist eye.

In Chapter IV, Dr. Tankanag presents a method for the analysis of peripheral blood flow oscillations is developed, which makes use of laser Doppler flowmetry (LDF) data and is based on continuous wavelet transform and the theory of adaptive wavelets.

In Chapter V, Dr. Amato introduces us to applications of Monte Carlo methods in medical physics, specifically in radiology, nuclear medicine and external beam radiotherapy, and in chapter VI he examines internal dosimetry in nuclear medicine.

In Chapter VII, Dr. Bogner describes methods that may reduce to a great extent errors and uncertainties in modern optimized radiotherapy with photons and protons.

In Chapter VII, Drs. Jiahui Wang and Qiang Li review the current status on computer-aided detection schemes of interstitial lung diseases in chest radiography and computed tomography

In Chapter IX, Dr. Balcerzyk explains essential issues of positron emission tomography of small animals in preclinical applications for engineers, radiochemists and biologists in a language comprehensible to the other, as many issues addressed by members of one field are difficult to comprehend by the members of another.

In: Medical Physics
Editor: Marcin Balcerzyk
ISBN: 978-1-62257-590-9

Chapter I

Mass Spectrometry: Medical Applications

Monica Culea
Babes-Bolyai University, Cluj-Napoca, Romania

Abstract

Mass spectrometry, including combined gas chromatography-mass spectrometry (GC-MS) is the most widely applied technology in metabolomics, as it provides a blend of rapid, sensitive and selective qualitative and quantitative analyses with the ability to identify metabolites. The aim of this work is to present some of our research results in medical applications by using GC-MS, especially the achievement of non-invasive or minimally invasive advanced methods of diagnosis and treatment of infants and children. The following subjects will be considered: Development of minimally invasive diagnosis method of the inborn metabolic error diseases (phenylketonuria (PKU) and maple syrup urine diseases (MSUD) and the associated biomarkers; Development and validation of diagnosis method for cirrhosis and hepatitis by SIM-GC/MS (the caffeine test); identification and selection of biomarkers for diagnosis of some diseases from breathed air; quantitative determination of anesthetics in surgery rooms; applications of stable isotope tracer in human nutrition research and metabolic studies; non-invasive theophylline determination by using isotope dilution mass spectrometric techniques.

Introduction

During the last years, a new array of analytical methodologies and technologies was introduced, namely, the scientific field of metabolomics, related to the analysis of microbial plant and animal metabolomes. The metabolomics experiment (including sampling and sample preparation, instrumental analysis, data processing and interpretation) provides unique challenges to fulfill the goal of improving the current status of biological information related to the metabolome and more generally functional genomics. The metabolome was defined as the qualitative and quantitative collection of all low molecular weight molecules (metabolites) present in a cell that are participants in general metabolic reactions and are required for the maintenance, growth and normal function of a cell. The size of the metabolome may cover a relatively large field, depending on the organism studied, the human metabolome being of the larger size.

The metabolome is extended over a wide concentration domain (from pmol to mmol). Metabolites constitute a diverse set of atomic arrangements when compared to the proteome (arrangement of 20 amino acids) and transcriptome (arrangement of four nucleotide bases bonded with sugar and phosphate backbone) and contains compounds with a wide variation of physical and chemical properties. Strategies for metabolomics analysis are: (1) metabolomics—identification and quantitation of all metabolites in a biological system, (2) metabolite profiling—identification and quantitation of a selected number of pre-defined metabolites (sample preparation and instrumentation are employed to isolate the compounds of interest from the possible matrix, followed by chromatographic separation and detection with MS), (3) metabolic fingerprinting—rapid, global analysis of samples, (4) metabolite target analysis—qualitative and quantitative analysis of one or a few metabolites related to a specific metabolic reaction and (5) metabonomics—changes in endogenous metabolite levels in tissues and biological fluids that result from disease or therapeutic treatments. It is estimated that metabolomics experiments cost two to three times less than proteomic and transcriptomic experiments. In recent years, there have been different applications developed by mass spectrometry (MS), gas chromatography – mass spectrometry (GC-MS) and nuclear magnetic resonance (NMR) for the determination of metabolic biomarkers that change as an indicator of the presence of a disease or in response to drug-based intervention, and determination of the effect of biochemical or environmental stresses on plants or microbes, which include genetically modified (GM) plants, bacterial characterizations, human health assessments and metabolic engineering [1].

Validation of an analytical method and instrument is important to assess their precision, accuracy and applicability. The time and method of sampling can greatly influence the reproducibility of the analytical sample.

Diurnal and dietary influences can have major effects on the composition of the metabolome. The storage of samples is important, as the continued freezing/thawing of samples can be detrimental to stability and composition. All these influence the precision, accuracy and reproducibility of results. Mass spectrometry, including GC-MS, is the most widely applied technology in metabolomics due to the fact that it provides a blend of rapid, sensitive and selective qualitative and quantitative analyses with the ability to identify metabolites.

Mass spectrometers operate by ion formation, separation of ions according to their mass-to-charge (m/z) ratio and detection of separated ions. GC-MS is a combined system where volatile and thermally stable compounds are first separated by GC and then eluting compounds are detected traditionally by electron-impact mass spectrometers. In metabolomics, GC-MS was described as the basic method. Volatile, low-mass weight metabolites (including breath and plant volatiles) can be sampled and subsequently analyzed directly.

The majority of analyzed metabolites require chemical derivatization at room or elevated temperatures to provide volatility and thermal stability prior to analysis. Sometimes the derivatization is done in two stages, as in amino acids case, by esterification at the C terminal and trifluoroacetylation at the N terminal.

Chromatograms are complex, containing hundreds of metabolite peaks. In addition, the chromatograms are further complicated by multiple derivatization products. Quantification at nmol or pmol levels is provided by either external calibration or response ratio (peak area of metabolite/peak area of internal standard).

Calibration is labour intensive, and not all metabolites are commercially available to enable preparation of standard solutions. Consequently, metabolite identification is provided by matching the retention time or retention index and mass spectrum of the sample peak with those of a pure compound previously analyzed under identical instrumental conditions. The mass spectra libraries (Wiley, NIST, EPA) can be successful for identification, but commercially available mass spectral databases do not contain all metabolites expected in a metabolic reaction. GC-MS applications are numerous, since GC-MS represents a mature analytical technique.

The first applications of GC-MS in metabolomics can be traced to urinary screening procedures to indicate the presence of diseases related to organic acidemias. Diseases such as cancer, diabetes, cardiovascular and neurodegenerative diseases, as well as the age-related illnesses, are caused by chemical modification of proteins, as a result of oxidative stress [2] mediated by reactive oxygen and nitrogen species. These diseases are characterized by formation of oxidized nucleotides and amino acids and by chemical modifications of proteins, sugars and lipids, known respectively as advanced glycoxidation and lipoxidation, excreted in urine as a result of normal turnover processes.

The chemical modifications of biomolecules in the body constitute the chemome, and the analysis of these nonenzymatic products (glycohemoglobin, isoprostanes, malondialdehyde, nitro- and ortho-tyrosine) can provide a unique insight into health, disease, risk for disease, and response to therapy. Such changes also appear with age; but the rate of their accumulation is accelerated in chronic disease. The latest generation of instruments provides high sensitivity and specificity for detection and quantification of trace analytes and allows for simultaneous analysis of a wide range of analytes as the measurement of amino acids, carboxylic acids or drugs in biological fluids. The biomarkers of oxidative stress [3] are compounds useful in finding disease in chemom global assessment in cardiovascular disease, metabolic diseases [4, 5], and cancer. Nutrients in the diet influence life, health, fertility [6, 7].

Transmembranar transport pathways elucidation contributes to understanding of body and is of great interest for health [8, 9]. The pharmacokinetics of drug are useful in serious disease diagnosis [5].

Our aim is to present results as: Preparation of amino acids and drugs standard samples; Tests of extraction, derivatization and qualitative determination of biomarkers: amino acid (AA), drugs in biofluids using standards; Methods validation of the extraction and quantitative analysis by selected ion monitoring (SIM)-GC-MS and GC-MS using isotopic labeled internal standard; Clinical monitoring of biomarkers (amino acids, drugs) for optimal treatment or diagnosis of diseases: asthma, genetic diseases, metabolic diseases, cirrhosis.

1. Development of Minimally Invasive Diagnosis of the Inborn Metabolic Diseases Phenylketonuria (PKU) and Maple Syrup Urine Diseases and the Associated Biomarkers

Introduction

Mass spectrometry (MS) is used in the development of new methods to study, diagnose and monitor treatment of patients with rare and genetic metabolic disease. Each year, millions of babies in the United States have dried blood spots analyzed through newborn screening programs. This screening is intended to detect inborn disorders that can result in early mortality or lifelong disabilities such as metabolic disorders (e.g., phenylketonuria (PKU)), hematologic disorders (e.g., sickle cell disease), and endocrinopathies (e.g., congenital hypothyroidism). Gas chromatography coupled with mass spectrometry (GC-MS) is typically used to quantitate and identify primarily non-polar metabolites that are less than 1 kDa in mass. GC-MS demonstrate that it is an indispensable method for diagnosing inborn error of metabolism and is widely recognized for its effectiveness in related fields [10, 11, 14-26, 31-34]. Recent papers report that the number of diseases in humans known to be attributable to inherited point defects in metabolism exceeds 500. While the diseases are individually rare, they collectively account for a significant proportion of illness, particularly in children. Metabolic diseases arise from inherited errors in enzymes in any number of pathways, but the symptoms are inevitably due to the buildup of a biosynthetic precursor or product that the body cannot excrete. The first few days of a newborn's life can be critical to his future well-being, for it may be the only point at which physicians have a chance to tell if something is wrong and treat it, avoiding physical and mental retardation or even death. The signs of disease in patients with inborn errors of metabolism are the result of metabolic disturbances caused by deficiency of some catalytic or transport protein. Accumulation of the substrate of a mutant enzyme is an important cause of disease in many inborn errors of metabolism, particularly those involving strictly degradative processes. PKU was shown to be caused by a recessively inherited point defect in metabolism, in the conversion of phenylalanine to tyrosine in the liver due to phenylalanine hydroxylase deficiency, which causes an increase of phenylalanine in plasma and a decreased tyrosine, a specific pattern of blood amino acids. MSUD is an inherited metabolic disease due to a deficiency in branched chain ketoacid dehydrogenase that leads to the accumulation of branched chain amino acids (BCAA), leucine, isoleucine, and valine, and their keto-acid derivatives. Diagnostic tests such as that for phenylketonuria, bacterial inhibition assay (BIA), which assays the inhibition of bacterial growth due to the presence of phenylalanine, look for specific markers indirectly and do not quantitatively assay the problem-causing metabolite itself [16, 17].

Mass spectrometry technique demonstrates that it is an indispensable method for diagnosing inborn error of metabolism and is widely recognized for its effectiveness in related fields. An alternative method was developed, without derivatization and chromatographic separation, by using electrospray mass spectrometry, ESI-MS-MS. Multiple reaction monitoring (MRM) was used instead of constant neutral loss scanning. The method was optimized for detection of amino acids, and quantitation was based on isotope-labeled internal standards [30]. Urease pretreatment-gas chromatography-mass spectrometry (UP-GC-MS) has become a valuable tool in the field of metabolome research, including analysis of inborn errors of metabolism and acquired metabolic disturbances secondary to nutrition or drugs [33].The methods use a drop of blood blotted onto a sample card; an aliquot is extracted for analysis, and peak profile is used to diagnose metabolic errors.

The present review presents the most recent methods for newborn diagnosis by using mass spectrometry technique.

Experimental

Extraction Procedure

Newborn testing by MS starts with the same steps used in the more traditional methods. A drop of blood is blotted onto a sample card, and an aliquot is extracted for analysis by MS. The small amount of blood is taken from a newborn within the first one to two postnatal days and is absorbed onto a piece of filter paper (Figure 1.1).

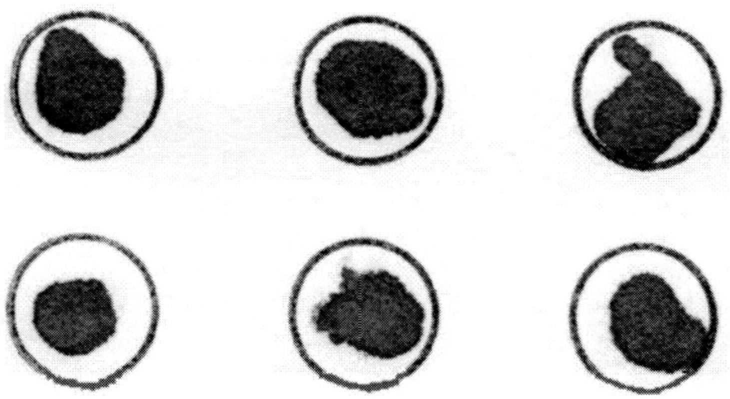

Figure 1.1. The drops of blood are blotted onto a sample card and an aliquot is extracted for analysis by MS or GC-MS. The peak profile is used to diagnose metabolic errors.

Table 1.1. Extraction procedures for amino acids from blood spot

Amino acids extraction	AA derivatization	
200 μl methanol/HCl 0.1% in a screw cap vial, 1h at 4°C or by sonication for 1 min. 100 μl of the extract placed in a vial and addition of the internal standard and removing of the reagent in a nitrogen stream	1. esterification:500μl butanol: acetyl chloride, 4:1 (v/v), 30 min, 100 °C. The excess reagent was removed with a stream of nitrogen	2.acetylation 100 μl trifluoacetic anhydride (TFAA) at 100 °C for 30 min, cool, dry, 1ml ethyl acetate

The small disk is punched out from the spot and extracted with various solvents. For example, a sample might first be immersed for several hours in a methanol solution containing isotopically labeled standards. After drying, the sample is derivatized in acidic 1-

butanol, washed in hexane, and dried before adding the solvent for GC-MS or MS analysis. The peak profile is used to diagnose metabolic errors [15, 16].

GC-MS Methods

GC-MS is a highly sensitive and specific technique used in organic analysis, in metabolism research and disease diagnostics. GC-MS enables the identification of small molecules as amino acids, fatty acids, organic acids in biofluids as blood and urine, providing diagnostic information in inherited diseases, including metabolic disorders [10, 11, 14-26, 31-34]. In the selective ion monitoring (SIM) mode, using a few selected ions, the sensitivity is increased by one or two orders of magnitude. SIM-GC/MS is very useful in quantitative work and is usually achieved by using isotope labeled internal standards. Quantitation is performed by addition of known amounts of internal standards to the sample before extraction. The method will compensate for sample losses in the clean-up stage, assuming that the losses of the standard are identical with those of the analyte.

A Trace DSQ ThermoFinnigan quadruple mass spectrometer in the EI mode coupled with a Trace GC was used. The capillary column Rtx-5MS was of 30 m length x 0.25 mm, 0.25µm film thickness, by using a temperature program from 50°C (1 min), then increased to 310°C, at 20 ° C/min, in the selected ion monitoring mode (SIM). Helium (99.9995%) carrier gas had a flow rate of 1 ml/min. The qualitative analysis was carried out in the mass range 50-500 a.m.u. Quantitative analysis was performed in the SIM mode by using the ions selected from the trifluoroacetyl butyl ester derivatives mass spectra: m/z 168 for valine, m/z 182 for leucine, m/z 166 for proline, m/z 91 and 148 for phenylalanine, m/z 203, 260, 316 (Figure 1.2) for tyrosine. Twenty-five µg/ml of the ^{15}N-isoleucine (^{15}N-Ile, m/z 183) internal standard was added at each sample. For Val, Leu and Pro amino acids, the important ions selected in the SIM experiment from the trifluoroacetyl butyl esters derivatives mass spectra correspond to the loss of butyl ester from the molecular ion $[M - COOC_4H_9]^+$[16].

Derivatization: Amino acids in blood samples or standard samples were derivatized as trifluoroacetyl butyl ester derivative [15-17] or trimethylsilyl derivatives [18]. Derivatization as trifluoroacetyl butyl ester derivative was followed in two steps, in screw-cap vials. Dry samples were esterified with 100 µl butanol: acetyl chloride, 4:1, v/v) for 30 min at 100 °C. The excess reagent was removed with a stream of nitrogen. The amino group was acetylated with 100 µl trifluoacetic anhydride (TFAA) at 100 °C for 30 min. After cooling, the excess reagent was removed under nitrogen at ice temperature and ethyl acetate was added (Table 1.1). The method is useful for diagnosis of metabolic diseases as PKU by determination of phenylalanine (Phe) and tyrosine (Tyr) in blood and MSUD by determination of valine (Val), leucine (Leu) and proline (Pro) in blood. The analysis of the five amino acids in blood samples by GC-MS is useful in the diagnosis of the both diseases.

Mass Spectrometry Method

MS/MS technology enables improvements and consolidation of metabolic screening methods to detect amino acid disorders (e.g., PKU, MSUD and homocystinuria) among newborns [11, 12, 27-29, 35]. MS/MS expands the metabolic disorder screening panel (i.e., the number of disorders that can be detected) by incorporating an acylcarnitine profile, which enables detection of fatty acid oxidation disorders (e.g., medium-chain acyl-CoA dehydrogenase [MCAD] deficiency) [32] and other organic acid disorders. Finally, MS/MS can reliably analyze approximately 20 metabolites in one short-duration run (~2 minutes) and provide a comprehensive assessment from a single blood-spot specimen.

The important element for the performance of a mass spectrometer is the mass analyzer. The most commonly used are the quadrupole, quadrupole ion trap, time-of-flight, and Fourier transform ion cyclotron resonance (FTMS) analysers. Quadrupole (Q) mass analyzers are used frequently since they tolerate relatively high pressures, have the capability of analyzing up to an m/z of 4000, and are relatively low-cost instruments. Tandem mass analysis places three quadrupoles in series, each with a separate function. Thus, the first quadrupole is used to scan m/z range and select an ion of interest. The second quadrupole, the collision cell, transmits the ions while introducing a collision gas, into the flight path of the selected ion. The third quadrupole analyzes the fragment ions generated in the collision cell. Quadrupole ion traps are useful in isolating a single ion species by ejecting all others from the trap, and the isolated ions can be fragmented then by collisional activation, and the fragments can be detected. The advantage of quadrupole ion traps is that multiple collision-induced dissociation experiments can be performed without having multiple analyzers. Other advantages of Q analyzers are their compact size and their ability to trap and accumulate ions to provide a better ion signal. The linear time-of-flight (TOF) is the simplest mass analyzer, with a resolution above 5,000, fast scanning capabilities, mass range up to m/z ~10000, and an accuracy in the order of 5 ppm.

Table 1.2. Metabolic disorders detectable in newborns, 1–5 days aged, by using tandem mass spectrometry

Disorder	*Metabolic indicator*
Amino Acids	
Phenylketonuria	Phe, Tyr
Maple syrup urine disease	Leu/Ile, Val
Homocystinuria (cystathione synthase deficiency)	Met
Hypermethioninemia	Met
Citrullinemia	Cit
Argininosuccinic aciduria	Cit
Tyrosinemia, type I	Tyr
Fatty Acids	
Medium-chain acyl-CoA dehydrogenase deficiency	C8, C10, C10:1, C6
Very-long-chain acyl-CoA dehydrogenase deficiency	C14:1, C14, C16
Short-chain acyl-CoA dehydrogenase deficiency	C4
Multiple acyl-CoA dehydrogenase deficiency	C4, C5, C8:1, C8, C12, C14, C16, C5DC
Carnitine palmitoyl transferase deficiency	C16, C18:1, C18
Carnitine/acylcarnitine translocase defect	C16, C18:1, C18
Long-chain hydroxy acyl-CoA dehydrogenase deficiency	C16OH, C18:1OH, C18OH
Trifunctional protein deficiency	C16OH, C18:1OH, C18OH
Organic Acids	
Glutaric acidemia, type I	C5DC
Propionic acidemia	C3
Methylmalonic acidemia	C3
Isovaleric acidemia	C5
3-hydroxy-3-methylglutaryl CoA lyase deficiency	C5OH
3-methylcrotonyl CoA carboxylase deficiency	C5OH

Table 1.3. Metabolic disorders detectable in patients aged >5 days by using tandem mass spectrometry

Disorder	Primary metabolic indicator
Argininemia	Arg
Nonketotic hyperglycinemia	Gly
Tyrosinemia, type II	Tyr
Hyperammonemia, hyperornithinemia, homocitrullinuria syndrome	Orn, HomoCit
5-oxoprolinuria	5-Oxopro
Carnitine palmitoyl transferase, type I deficiency	Free CN
Isobutyrl CoA dehydrogenase deficiency	C4
Mitochondrial acetoacetyl CoA thiolase deficiency	C5:1,C5OH
Malonic aciduria	C3-DC
2-methylbutyrl CoA dehydrogenase deficiency	C5

FTMS offers high resolution, the ability to perform multiple collision experiments (MSn) and high-accuracy fragment masses, respectively, high resolution. FTMS coupled to a wide variety of ionization sources, including MALDI (Matrix Assisted Laser Desorption Ionization), ESI (Electrospray Ionization), APCI (Atmosferic Pressure Chemical Ionization), and EI (Electron Impact).

A PE Sciex API 2000 (Perkin-Elmer Sciex, Toronto, Canada) benchtop triple-quadrupole mass spectrometer equipped with a Turbo IonSpray source was used [30]. The probe was operated at a sprayer voltage of 5 kV. The electrospray source temperature was set to 100°C. Curtain gas back pressure of 0.14MPa (20 psi), nebulizer gas of 0.42MPa (60 psi), and turbo gas of 0.21MPa (30 psi) were applied. A 20-mL sampling loop was fitted to the injector; flow rate was 50 mL/min using methanol solvent. Peak intensities were determined by averaging the plateaus of the total ion currents. The multiple reaction monitoring (MRM) technique was applied for tandem mass spectrometric measurements, such that each metabolite had its individual precursor and product ion settings. Specific fragments were determined for 23 L-amino acids from product ion scans obtained using continuous flow sample infusion with a flow rate of 5 mL/min. All the source parameters, gas flow and collision energy values were optimized for each component in the MRM experiments.

Tables 1.2 and 1.3 present the metabolic indicator (biomarker) determined in different metabolic disorders in patients at different ages.

Results and Discussion

The GC-MS was used in the SIM mode in the range 50-500 a.m.u. for qualitative determination of the amino acids to obtain the mass spectrum of each amino acid of interest. The mass spectra of the amino acids used for PKU profiling determination are presented in Figure1.2.

Validation of the methods was studied. Linearity of the method was calculated by representing the ratio of the selected ion peak area for each amino acid and the internal standard versus the amino acid standard concentrations (in μg/ml). The regression curves were obtained by injecting standard solutions containing amino acids in the range 1- 40 μg/ml with or without addition of internal standard.

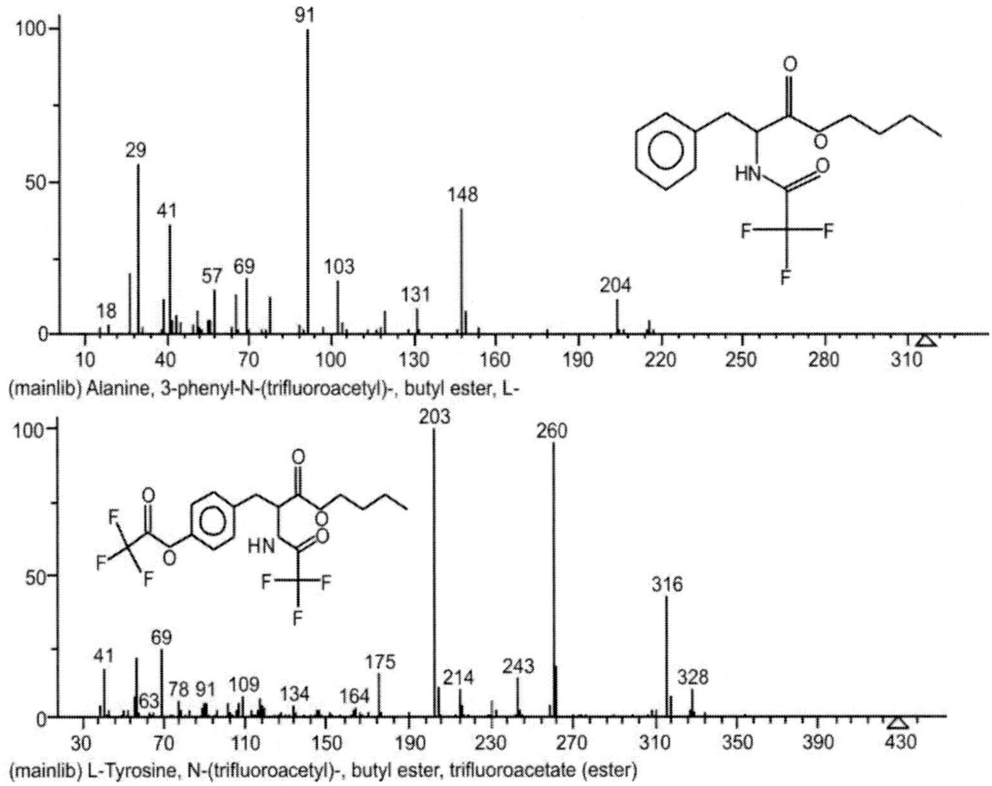

Figure 1.2. The mass spectra of the phenylalanine and tyrosine as trifluoroacetic butyl ester derivatives used for PKU diagnosis.

The regression curves gave good regression coefficients (r >0.998) for the amino acids of interest. The precision and the accuracy gave good R.S.D. values (<20%). The limit of detection (L.O.D.) was lower than 0.1 μg /ml [15-18]. Our results were obtained for quantitative values of the five amino acids of blood spots of 20 μL. The precision and the accuracy gave R.S.D. values ranged between 9–12.9% for 30 μg/mL and 6.7–18.6% for 40 μg/mL. Accuracy values were lower than 5.5% for 40 μg/mL (n = 4). The limit of detection (L.O.D.) was lower than 0.1 μg /mL. The values for the amino acids in 53 control subjects (n =53) showed higher coefficients of variation (R.S.D. is higher due to intraindividual differences), but in the case of the same blood sample (n = 6 spots), precision gave R.S.D. lower than 10% [15, 16].

Table 1.4. Plasma amino acid concentrations mean values in a newborn and a child (control)

Plasma	Newborn (μM/l)[23,37]	Newborn (μg/ml)[23,37]	Child (μM/l) [15]	Child (μg/ml) [15]
Val	136.75	16	28.53	3.34
Leu	72.52	9.5	64.14	7.50
Pro	185.22	21.3	163.88	19.17
Phe	78.79	13	88.06	10.30
Tyr	69.61	12.6	69.65	8.15

Table 1.5. Comparison of the amino acids values determined by different methods (in μmol/L) [15-18, 30]

Blood	SIM-GC-MS, TFA-butyl ester derivatives [15,16], PKU patient	SIM-GC-MS, TFA-butyl ester derivatives [17], PKU patient	SIM-GC-MS, TFA-butyl ester derivatives [17], MSUD patient	SIM-GC-MS, Trimethylsilyl derivatives [18], PKU patient	MS/MS without derivatization [30], control
Val	198.50	107	356	-	461.2
Leu	197.29	161	667	-	197.29
Pro	222.98	212	452	-	201.3
Phe	864.73	1023	113	673	157.4
Tyr	198.41	208	203	186	138.9
Phe/Tyr	4.36	4.92	0.54	3.82	1.13

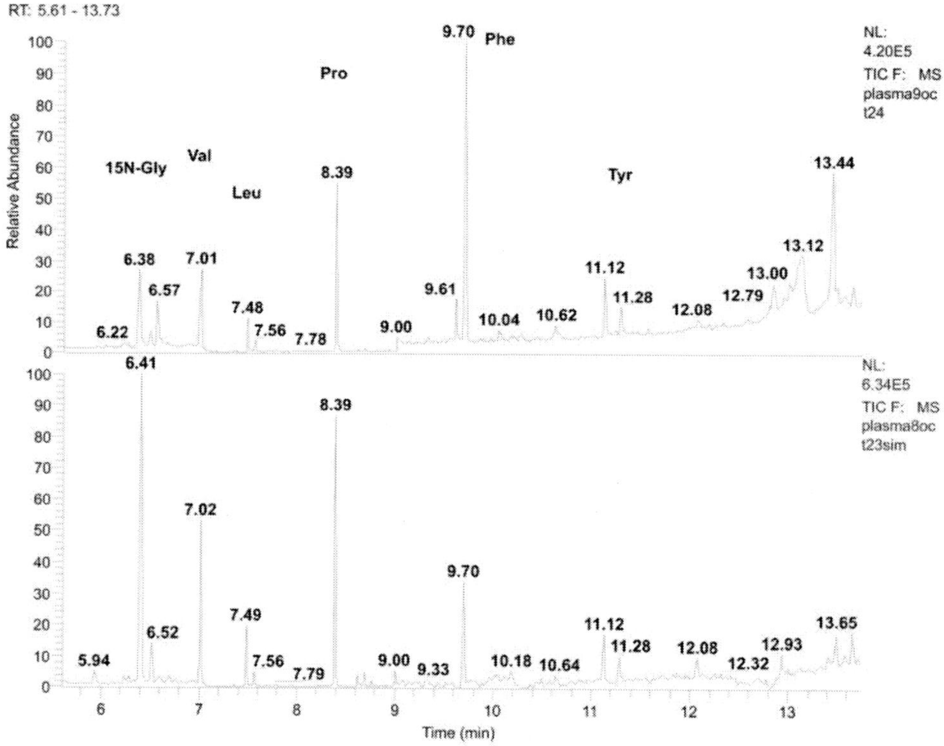

Figure 1.3. The fingerprint chromatograms in SIM-GC-MS for a PKU patient present the differences between the first and the third month of treatment [16]. The retention times for the amino acids: ^{15}N-Gly at 6.40 min, Val at 7.02 min, Leu at 7.50 min, Pro at 8.40 min, Phe at 9.70 min and Tyr at 11.28 min.

Good values for precision and accuracy, R.S.D. values (<20%), were also obtained by other authors [17, 18]. The amino acid values are different among adults, children and newborn. Table 1.4 presents the comparison of the results obtained for the five amino acids average value in newborn blood samples [37] and the values obtained by blood spot analysis for a child (n = 3, our results) [15].

PKU diagnosis could be tested by calculating the ratio Phe/Tyr using 20 μL of blood spots. A comparison of two chromatograms (Figure 1.3, our results) demonstrates the benefit of the treatment in the case of a PKU child [15, 16]. The first fingerprint chromatogram (top) presents higher intensity of the phenylalanine peak in the case of a patient presenting PKU, and after three months of treatment (second chromatogram, down) the peak of Phe is significantly decreased, the ratio of Phe/Tyr decreased from 2.30 at 1.02.

Diagnosis of MSUD disease is assessed by the determination of Val, Leu and Pro and by calculating the ratio between aliphatic and aromatic amino acids in the blood samples.

Table 1.5 shows the comparison of the results for the amino acids values obtained by blood spot analysis in PKU patients by different extraction, derivatization or analysis methods. The amino acids determined by different methods and different authors show acceptable interindividual differences [15-18, 30]. The last column contains amino acid values for a control subject. The ratio of Phe/Tyr in the PKU positive blood was more than 2.0, while in controls it was less than 1.2. In MUSD blood samples (column 3), aliphatic amino acids are present at higher levels than aromatic amino acids, and the ratio is >2. The third column shows higher values for Leu and Pro, but Phe and Tyr are normal.

Metabolite Profiling

Metabolite profiling can be used to determine drug efficiency, toxicity or disease states. Alterations of the metabolites in diseases or drug intervention, the changes in gene and protein expressions or enzyme activity can be explored [36]. Many clinical laboratories around the world use GC-MS for diagnosis of different metabolic diseases, particularly those disorders of organic acids. The advantages of mass spectrometry include a wide dynamic range, the ability to observe a diverse number of molecular species, and reproducible quantitative analysis. These attributes are important in addressing the issue of metabolite profiling, as the dynamic range easily exceeds nine orders of magnitude in biofluids, and the diversity of species ranges from simple amino acids to lipids to complex carbohydrates.

The goals of the application of mass spectrometry range from basic biochemistry to clinical biomarker discovery with challenges in generating a comprehensive profile, data analysis, and structurally characterizing physiologically important metabolites.

Blood samples from millions of neonates are tested routinely by mass spectrometry as a diagnostic tool for inborn errors of metabolism. The BIA method is a semi-quantitative method used for PKU diagnosis, but it indicates a false positive response up to 5% [16, 17]. GC-MS is rapid, sensitive, selective, simple and less expensive, measuring several amino acids at once. The GC-MS methods are minimally invasive methods suitable for PKU diagnosis in neonatal blood samples, either by determination or screening of some amino acids, by using very small quantities of blood. The ratio of aliphatic amino acids to aromatic amino acids could indicate other disorders of the amino acid metabolism such as MSUD.

Conclusion

GC/MS is the most commonly used method for small molecule analysis, and it is still used today for the detection of many metabolic disorders. GC/MS enabled the identification of key small molecules – such as amino acids in biofluids, particularly in urine and blood. GC/MS method is a simple, inexpensive, easily operated and high-speed technique.

Addition of internal standard materials at the beginning of the extraction procedure allows monitoring the extraction and derivatization as well as instrument performance. The internal standard used increased the precision of the method [16, 18].

The disadvantages of using GC-MS in applications to metabolic profiling include sample preparation, derivatization to improve volatility, lengthy analysis time, type of molecule that could be analyzed. Polar or nonvolatile molecules could be analyzed by LC-MS with an electrospray interface and MS/MS techniques.

Mass spectrometry is already well established as a quantitative tool for small molecules, unlike its application to proteomics. Mass spectrometry also facilitates the identification of previously uncharacterized metabolites. Because MS-MS instrument is too expensive for low-budget hospitals, it is not widely applied to diagnosis of aminoacidemias including PKU and MSUD in developing countries. By comparison, the GC/MS methods are simple, inexpensive and easily operated. As PKU, many metabolic disorders can now be easily monitored and many drugs can be administered. The area of metabolite profiling is expanding rapidly, and the applications for this science range from basic biochemistry to clinical-biomarker discovery.

2. Gas-Chromatography-Mass Spectrometric Method for Liver Function Test in Children

Introduction

Gas chromatography-mass spectrometric (GC/MS) applications for measuring drugs are very important for purity control [38, 39], for pharmacokinetic studies [40, 41], in metabolic studies or clinical applications for treatment and diagnosis [43]. Caffeine is used for measuring the metabolic capacity of the liver but also as a marker of river waters pollution [44]. Fasting plasma or saliva caffeine concentrations have been suggested to be a simple guide to the severity of chronic liver disease [45-47]. Caffeine has the advantage of being well tolerated when administered orally, and the saliva level parallels the serum concentration, making a non-invasive test feasible [47, 48]. Because caffeine is metabolized by the hepatic P-450 cytochrome oxidase system, clearance of caffeine is an excellent quantitative test of hepatic function in human beings. The caffeine is demethylated under the influence of the hepatic microsomal cytocrome P-450 producing dimethylxanthines as theophylline and theobromine [46], so these xanthines should be avoided to be internal standards in quantitative work. Caffeine test consists of caffeine oral intake followed by measurements of the caffeine concentration or urine metabolites in blood, saliva, urine or of labelled CO_2 in the exhalation air. Metabolism of caffeine is decreased in patients with primary billiary cirrhosis and alcoholic liver cirrhosis. The caffeine clearance test was used also to assess the improvement in hepatic metabolic capacity after nutritional supplementation of branched-chain amino acids or after drug-induced hepatitis [49, 50]. The determination of caffeine clearance can serve as a useful parameter for the assessmentof hepatic functional reserve in hepatocellular carcinoma patients post treatment [51]. GC/MS studies of labelled xanthines have demonstrated that teophylline is converted to caffeine in premature human newborns by N-7 methylation. In adults, the inverse process exists wherein caffeine is demethylated to give theophylline [52]. The effects of liver disease on some pharmacokinetic parameters of caffeine as clearance and half-life time were usually studied by HPLC [53] and immunoassay methods [45] and sometimes by gas chromatography (GC) and gas chromatography-mass spectrometry (GC/MS) [54].

The development of a method is described to characterize some pharmacokinetic parameters as plasma clearance and half-life time of caffeine in children to test hepatic dysfunctions.

Experimental

^{15}N-theophylline, 74.2 atom % ^{15}N, labelled at the nitrogen in the position 7, synthesized in the National Institute for Research and Development for Isotopic and Molecular Technology Cluj-Napoca, was used as internal standard [43]. The purity was ascertained by infrared spectroscopy, mass spectrometry and melting point. Caffeine was administered p.o. in children with different hepatic dysfunctions. Caffeine, as a sterile caffeine sodium benzoate solution in water for injection use, contained 125 mg of caffeine and 125 mg of sodium benzoate/1 mL.

General Procedure

Extraction procedure: 1 mL of plasma containing caffeine was placed into a 5 mL screw-cap vial and 10 μL of internal standard ^{15}N-theophylline, 2 mL of the extraction solvent, chloroform: isopropanol 20:1 v/v and 0.5 g NaCl were added. After mechanical mixing for one min, the sample was centrifuged for three min. The organic layer (lower layer) was transferred to second vial and evaporated to dryness under a stream of argon. The residue was dissolved in 100 μL solvent and then 3 μL were injected into the GC.

Method validation: The method was validated in the range 0-20 μgmL^{-1}caffeine. Aliquots of distilled water containing known amounts of caffeine 3, 5, 10, 15, 20 μg mL^{-1}and 10 μg of ^{15}N-theophylline were taken through above procedure. Each sample was prepared in duplicate and measured twice. The regression curve, plotted as peak-area ratio of m/z 194-181 *versus* caffeine concentration, gave the following linearity parameters: slope 0.5207, intercept 0.1058, coefficient of correlation r = 0.98.

Precision gave RSD values lower than 5% for 5 μgmL^{-1}(n = 7) and lower than 3% for 3 μgmL^{-1}(n = 5). Accuracy showed values lower than 10% (Table 2.1). Each value was obtained as an average between two measurements of the same sample. The limit of detection was 0.1 μg mL^{-1} caffeine in blood sample, signal to noise ratio 4:1.

Population: Caffeine concentration measurements were performed in 33 hospitalized children suffering from hepatic dysfunctions and controls. Three different groups were studied: A, formed by 18 children with hepatitis aged 3-19 years old, B, consisting of five children with cirrhosis, aged between 5-12 years old, and C, ten children as control aged between 5-15 years old. The main dose was 4 mg/kg, p.o. for all groups. Blood samples were taken, at 0, 30 min, 1, 3, 6, 9 and 12 h. Blood samples were drawn into heparinized plastic tubes and immediately centrifuged. Plasma was stored at -20 °C. Written informed consents were obtained from each subject's parent prior to this study.

Calculation: Regression curves obtain by the GC/MS method in the SIM mode were used for pharmacokinetic parameters study. Caffeine elimination constant was calculated as follows:

$$k_{el} = (\ln C_1 - \ln C_2)/\Delta t$$

where C_1 = higher caffeine blood concentration, C_2 = lower caffeine blood concentration, Δt = the time elapsed between venous blood samples.

Two points caffeine clearance (Cl) was calculated as $Cl = k_{el} \times V_d$ and caffeine half-life as $t_{1/2} = \ln 2/k_{el}$, using a constant volume of distribution (V_d) of 0.6 L/kg body weight [45]. Clearance values calculated as dose/area under curve (AUC) were compared with the two-point values.

Detection method: A Hewlett Packard (Palo Alto, CA, USA) 5989B mass spectrometer coupled to a 5890 gas chromatograph were used in the following conditions: EI (electronic impact) mode, selected ion monitoring (SIM) mode, electron energy 70eV, electron emission 300 μA and ion source temperature 200 ºC. The GC/MS interface line was maintained at 280 °C and quadrupole analyser at 100 ºC. The gaschromatograph-mass spectrometer (GC/MS) assay used a HP-5MS fused silica capillary column, 30 m × 0.25 mm, 0.25 μm film-thickness, programmed from 200-250 °C at a rate of 10 °C/min, the flow rate 1 mL/min, with helium 5.5 as carrier gas. Injector temperature was 200 °C.

Table 2.1. Precision and accuracy of the method

Concentration added (μg ml^{-1})	n	Concentration measured (μg ml^{-1})	RSD (%)	Accuracy (%)
3	5	3.1	2.96	3.36
5	7	5.5	5.06	10

Retention time for caffeine and ^{15}N-theophylline, the internal standard, were 3.5 and 2.8 min, respectively. Three μL of sample were injected. The molecular ion m/z 194 for caffeine and the molecularion m/z 181 for the internal standard were monitored for quantitative analyses in SIM mode.

Results and Discussion

Caffeine clearance, measured in patients with cirrhosis and chronic hepatitis, was reduced, and half-life time was increased in children with liver disease as compared with control. The decreased metabolism observed in patients with various forms of liver disease was correlated to the disease status. The ion chromatogram in SIM mode for the molecular ions, the basic peaks for the both components, caffeine (m/z 194) and the internal standard (m/z 181), are presented in Figure 2.1. The mass spectrum of the caffeine is shown in Figure 2.2.

The mean values of the two pharmacokinetic parameters studied, Cl and $t_{1/2}$, for patients with hepatic diseases and control, show high differences especially between control and cirrhosis cases. Caffeine plasma concentrations at eight h time interval in hepatitis control and cirrhotic patients are presented in Figure 2.3. Figure 2.4 presents the plasma mean values of caffeine concentration in patients in comparison with the mean value of caffeine concentration obtained in control. The elimination curve of plasma caffeine obtained for a control subject is presented in Figure 2.5.

The method developed is simple, precise and rapid and useful in the analysis of xanthines. The isotopic labelled internal standard and SIM method used avoid the overlapping with different contaminants or metabolites. Good linearity, precision, accuracy and sensitivity were obtained in the range 0-20 μg/mL of the drug. Significant changes (student's paired t-test $p < 0.01$) were observed in caffeine metabolism in children with decompensate cirrhosis.

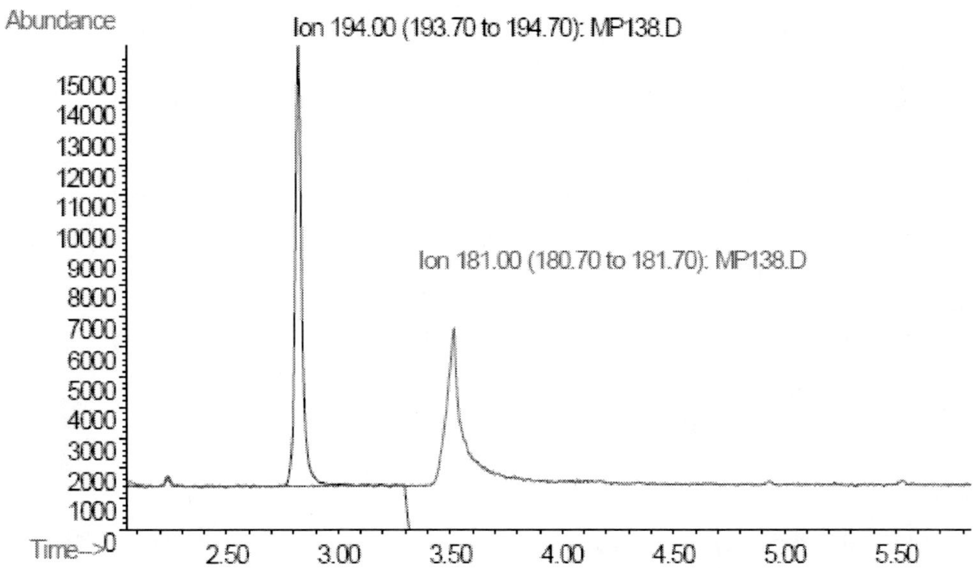

Figure 2.1. Caffeine separation chromatogram by GC/MS in the SIM mode; basic ions m/z 194 for caffeine and m/z 181 for the internal standard have been used.

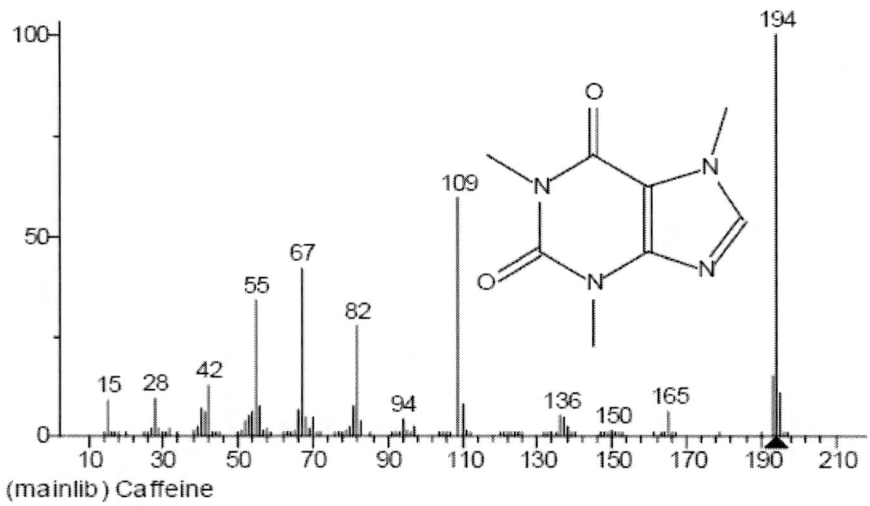

Figure 2.2. The mass spectrum of caffeine; M=194.

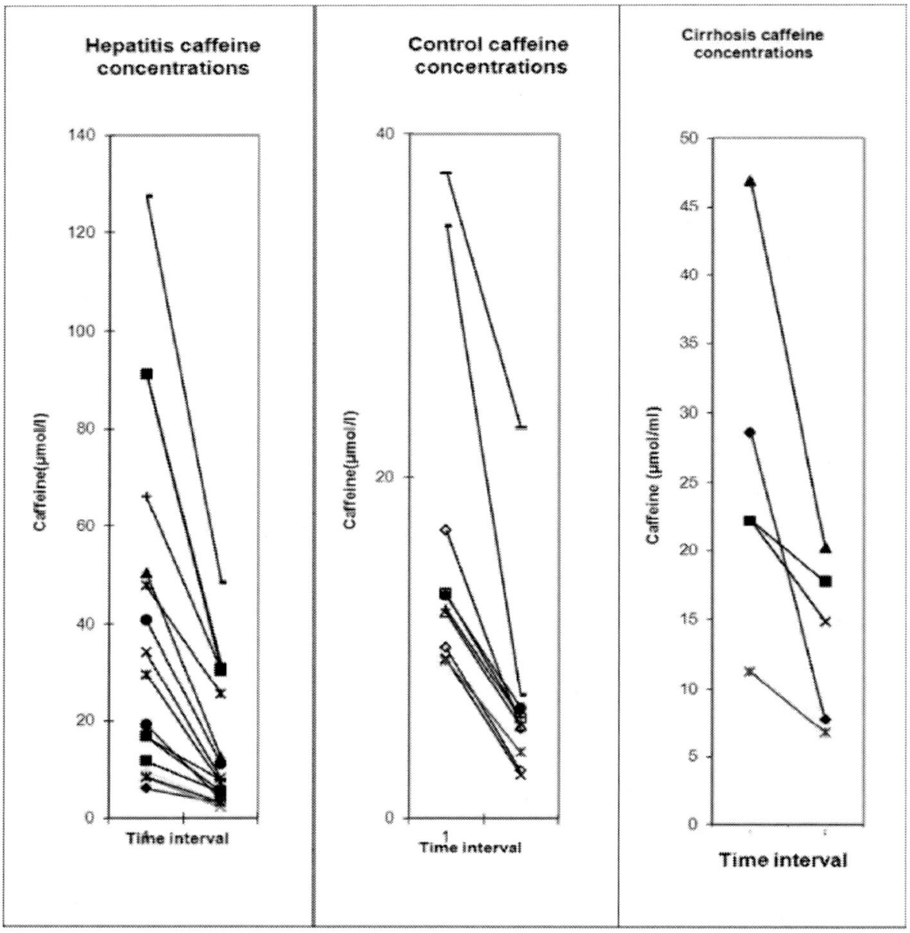

Figure 2.3. Caffeine plasma concentrations at time interval of 8 h in hepatitis, control and cirrhotic patients (two-point concentration in given subject is connected with a straight line).

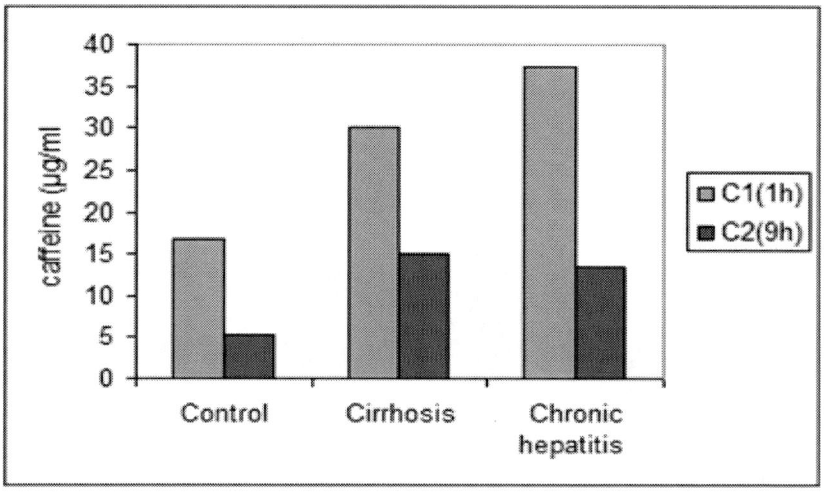

Figure 2.4. Mean level of caffeine after 1 and 9 h after dose in the blood of patients in comparison with control.

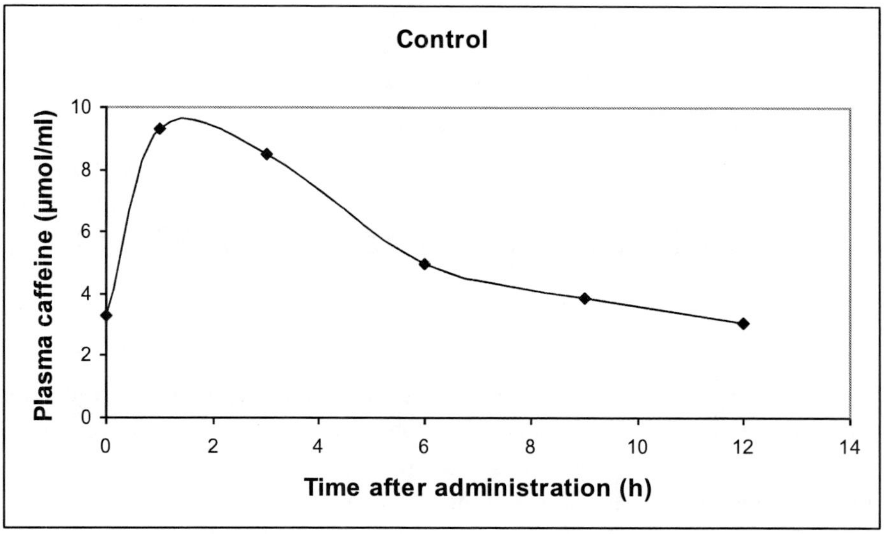

Figure 2.5. The elimination curve of caffeine in a control subject.

The clearance values of 0.74 ±0.49 mL $min^{-1}kg^{-1}$and half-life times of 14.73 ± 12.36 h are changed because of the reduction in "functioning hepatocyte mass." Patients with noncirrhotic liver disease showed intermediate values (Cl = 1.23 ± 0.45 mL $min^{-1}kg^{-1}$and t1/2 = 6.33 ± 2.17 h) but higher values of caffeine plasma concentrations especially in the first hour after dose (Figure 2.4).

The control levels for clearance and half-life time were 1.3 ± 0.4 mL $min^{-1}kg^{-1}$and $t_{1/2}$= 4.4 ± 1.9 h in the literature [46] and, respectively, 1.28 ± 0.31 mL $min^{-1}kg^{-1}$and $t_{1/2}$ = 5.73 ± 1.58 h (n = 10) in our data.

Plasma concentrations of caffeine were measured in eighteen patients with chronic hepatitis, five patients with cirrhosis and in ten healthy subjects after caffeine (4 mg/kg p.o.) loading. The total body clearance measured by two-point study (sampling times 1 and 9 h) and seven-point study (sampling times 0, 0.5, 1.0, 3.0, 6.0, 9.0, 12 h) were highly significantly correlated (r = 0.94, p less than 0.001).

The elimination half-life ($t_{1/2}$) of caffeine was significantly longer in cirrhotic patients than in the other two groups. The clearance was substantially reduced in cirrhotic patients. These findings suggest that caffeine pharmacokinetic parameters can be estimated using two-point blood sampling procedure and GC/MS determination, following a single load. The higher concentrations of caffeine observed in the first hour after caffeine loading in hepatitis compared with controls could be used for a rapid test for hepatitis when very precise and accurate methods are used.

Conclusion

The method developed is simple, precise and rapid and useful in the analysis of xanthines. The use of isotopic labelled internal standard avoids the overlapping with different contaminants. Good validation parameters were obtained in the range of interest.

Significant changes ($p < 0.01$) were observed in caffeine metabolism in children with decompensated cirrhosis. The clearance values in cirrhosis are changed because of the reduction in "functioning hepatocyte mass." Patients with noncirrhotic liver disease showed intermediate values but higher values of caffeine plasma concentrations

3. Leukotriene Quantitation by GC-MS

Introduction

Quantitation of eicosanoid levels in biological systems is important for understanding of their role in cell function and for diagnosis. Leukotrienes quantitation is limited by sensitivity of the method. Gas chromatography–mass spectrometry (GC-MS) provides a sensitive and specific method when proper derivatization method is used.

Bronchoscopy with bronchoalveolar lavage is the gold standard to assess airway inflammation, but invasiveness makes it unethical, especially in children. The exhaled leukotriene B4 (LKB_4) is elevated in asthmatic children compared with control. LKB_4 is a non-invasive marker for airway inflammation in asthmatic children [56, 57].

A preliminary work is described for GC-MS use for quantitative determination of LKB_4 in exhaled breath content.

Experimental

Gas chromatography was performed on a 5% phenyl methylpolysiloxane column (30 m x 0.25 mm I.D., 0.25 µm film thickness) operated in a suitable temperature program. Helium carrier gas was of 1ml/min. Ionization was performed by electron impact (EI) and detection in SIM mode. Arachidonic acid was used as internal standard.

Exhaled Breath Condensate Collection

A condensing chamber (Ecoscreen II, Germany) was used to obtain exhaled breath condense (EBC). 1.5 ml of EBC was obtained after 15 minutes of breath through a mouthpiece connected to the condenser. Samples were stored at -80°C before measurements.

Patients with asthma were recruited from Paediatric Clinic III of Cluj-Napoca. The preliminary study was for three patients with asthma and three controls.

GC-MS Method

LKB_4 and arachidonic acid, the internal standard, were analyzed by GC-MS after derivatization with ethylchloroformate and liquid-liquid extraction. A modified method was used [56]. Reagents were from Merck (Germany). Standards containing different concentration of LKB_4 and the same concentration of arachidonic acid (IS) were prepared in water. Deactivation of all glass surfaces, injector liner and vials, was made with 10% dichlorodimethylsilan in hexane for 30 minutes.

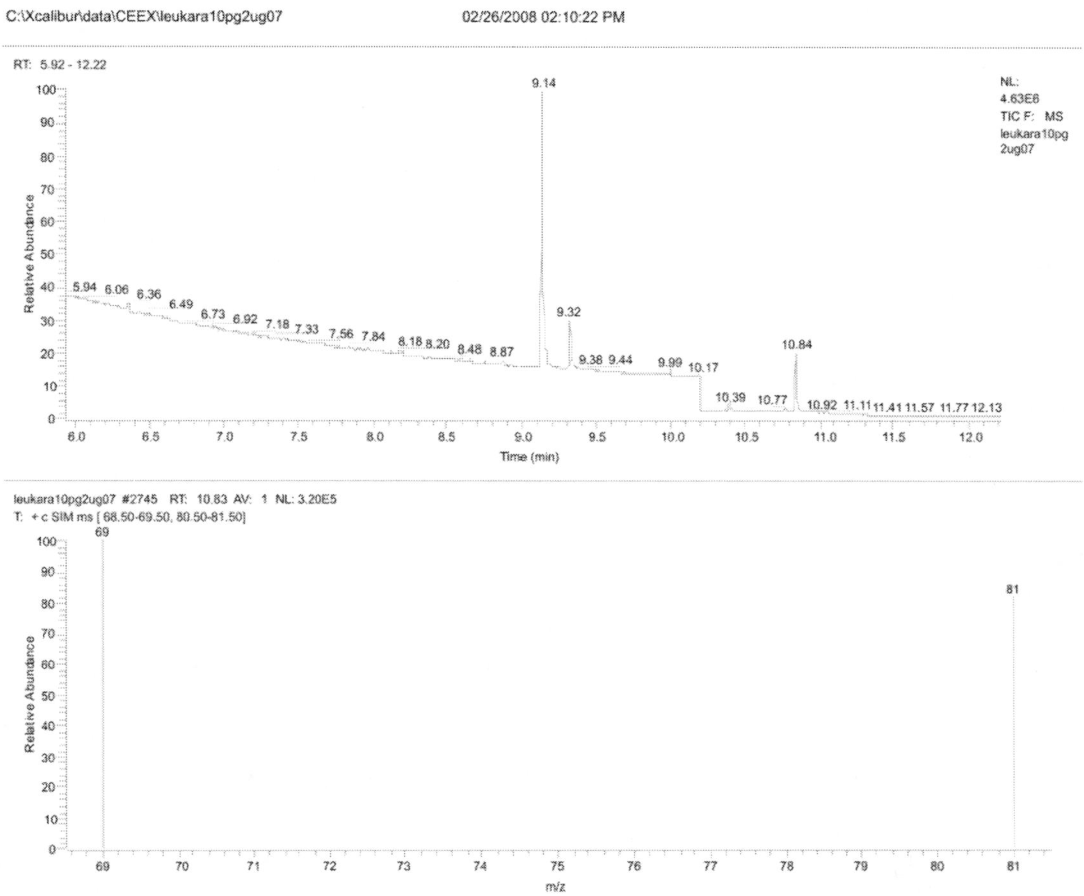

Figure 3.1. SIM chromatogram for arachidonic acid ethyl ester, the internal standard, elution time 9.33 min, and derivatized LKB_4 elution time 10.85 min. The ions m/z 69 and 81 are used in SIM mode for LKB_4. The distilled water (500 µl) was spiked with 1µg arachidonic acid and 10pg LKB_4.

Table 3.1. LKB_4 in EBC values in asthmatic and control children

Patient	LKB_4 (pg/ml)	RSD(%)
1	178.2	32
2	219.5	47
3	315	35
Controls (n=3)	78	36

Five hundred µl EBC, 1 µg arachidonic acid (IS), 20 µl ethylchloroformate and 40 µl pyridine were shaken few second for evaporating CO_2. 0.25 ml chloroform, 0.5 ml sodium carbonate-bicarbonate solution were shaken 30 seconds. After two minutes of phase equilibrium, 3 µl of the lower phase was injected into the fused silica capillary column in the split-less mode. Injector temperature was 280°C.

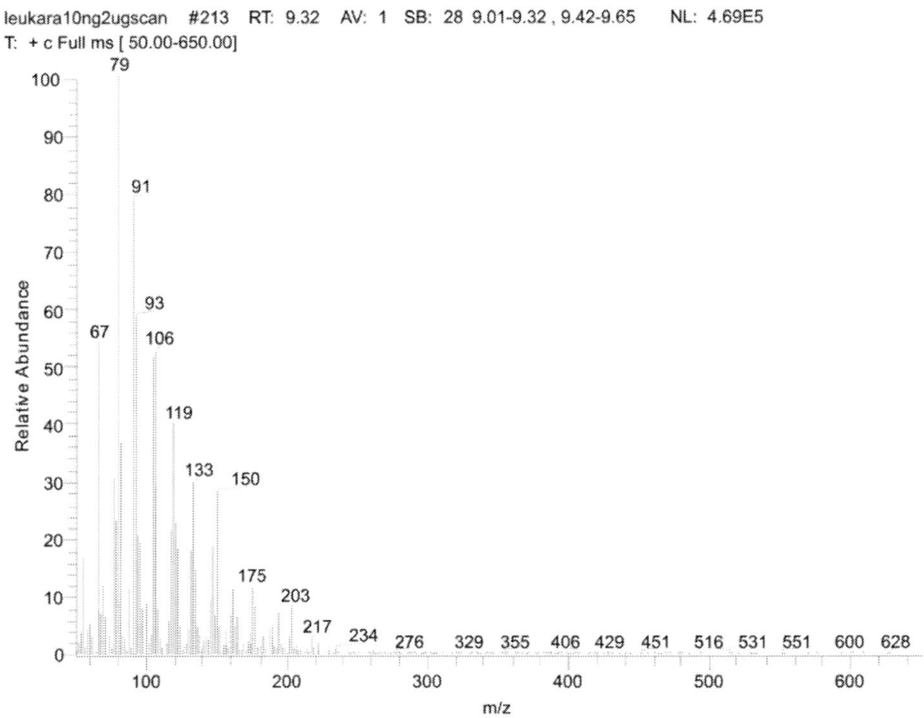

Figure 3.2. The mass spectrum of arachidonic acid ethyl ester (9.33 min), M=332.

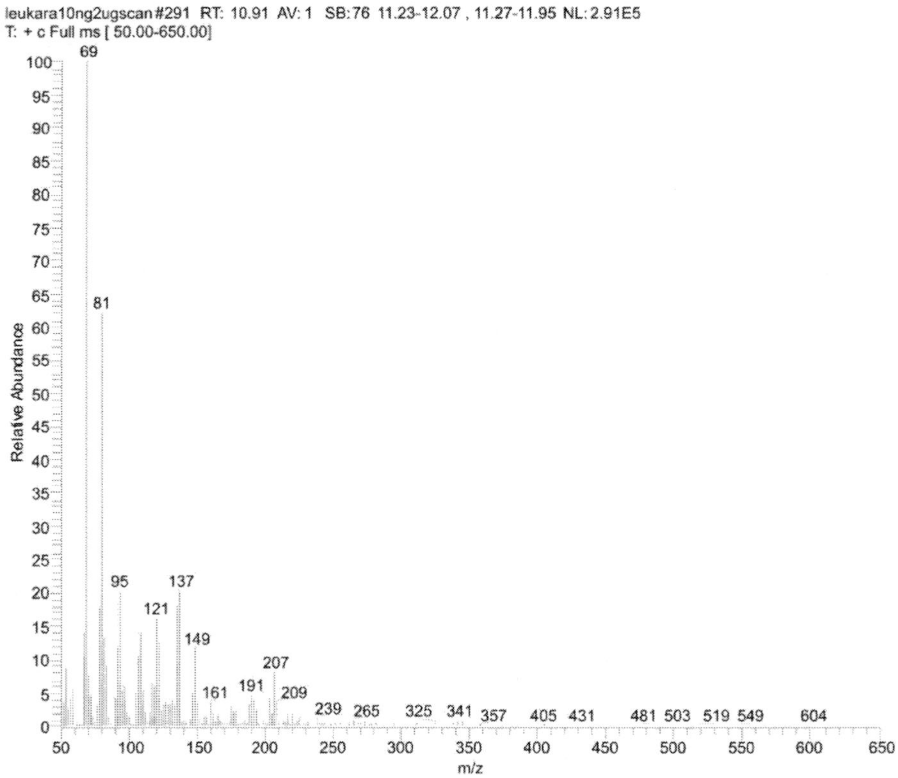

Figure 3.3. The mass spectrum of leukotriene B4 ethyl ester (10.85 min).

Gas chromatography was performed on a 5% phenyl methylpolysiloxane column (30 m x 0.25 mm I.D., 0.25 μm film thickness) operated in the following temperature program: initial temperature 80°C (1.5 min), then increased with a rate of 28 °C/min at 315 °C and kept five minutes. The ion current chromatogram for the two compounds is presented in Figure 3.1. The selected ions were m/z 79 for arachidonic acid derivative and m/z 69, 81 for LKB_4 derivative. The mass spectra for the two derivatized compounds are presented in the Figures 2 and 3.

Helium carrier gas flow was of 1ml/min. Ionization was performed by electron impact (EI) and detection in SIM mode.

Results and Discussion

The method provided good response linearity and precision (< 30% C.V.) in the 1-400pg/ml range, necessary to analyse leukotrene B4 in exhaled air. Arachidonic acid was used as internal standard because its structure is similar to leukotrienes, it is undetectable in EBC without addition of arachidonic acid and it is stable in aqueous solution.

The linearity of the method was studied in the range1-400 pg/ml. The regression curve obtained was: y=0.0027x + 0.4038, with a regression coefficient of r=0.99. The detection limit was of 1 pg/ml, at a signal-to-noise of five.

The method was applied for few cases of patients and controls. The different values of LKB_4 were significant higher in asthmatic children than in healthy children.

Measurements of exhaled LKB_4 are important for identifying the children with no respiratory symptoms but having inflammations that may require therapy.

Conclusion

The LKB_4 was detectable in exhaled breath condense. GC-MS non-invasive method showed significantly increased value for LKB_4 in children with asthma compared to healthy control.

4. Determination of Halothane Fumes in Operating Theatre

Introduction

The occupational exposure to anaesthetic gases is of great concern in the medical world. The air contamination was assessed in hospitals by using gas chromatography with mass detection or flow ionization and by infra-red spectroscopy or by using headspace sampling and solid phase microextraction (HS-SPME), followed by GC-MS analysis. The medical staff working in surgical wards, doctors and nurses, as well as people involved in the production, transport and storage of anaesthetics are professionally exposed to fumes released in their working environment [57-70]. The present study is devoted to the determination of halothane concentration in the indoor environment of hospital surgery rooms in order to get an evaluation of the professional risk. Measurements were performed in different surgery halls, by collecting samples of the air presumably contaminated with halothane fumes, during operation under narcosis, in each case.

The aim of the study was to evaluate the efficiency and accuracy of halothane determination by comparing two techniques of analyzing the air contaminants in surgery halls, by using a gas chromatographic technique with flame ionization detector FID (method 1) and a GC-MS coupling technique in the selected ionization mode SIM (method 2). Halothane is a usually well-tolerated inhaled anaesthetic, though sometimes complications may occur in patients [57, 59, 67] and in medical staff [58, 67-70]. The exposure to anaesthetic gases represents the main chemical risk for most people working in an operating theatre [58, 70]. There are various techniques used for the detection and analysis of halothane in the air [57, 61, 63, 68], based on photoacoustic effect [68], IR spectroscopy, gas chromatography (GC) with FID [62-64] or ECD detector [57]. Prior to the injection onto the chromatographic column, in most studies, halothane was collected on active charcoal, followed by liquid extraction in benzyl-alcohol [57] or in toluene [63].

Experimental

Extraction Procedure

Apparatus: In method 1, a Perkin Elmer Model 990 gas chromatograph equipped with a flame ionization detector (FID) was used, and the analytes were separated on a DB-5 (30 m x 0.25 mm, 0.5 μm film) fused-silica capillary column. The split ratio was 15:1, the injector and the detector temperature were kept at 250 ^{o}C. The column temperature was initially held at 40 ^{o}C, for two minutes and then increased to 170 ^{o}C at a rate of 30 ^{o}C/min. In method 2, the analytes were separated on a fused silica capillary column Rtx-5 MS (15 m x 0.25 mm, 0.25 μm film), kept at 30 ^{o}C (3 min), and then increased to 120 ^{o}C at 50 ^{o}C/min. The detector was a Thermo Finnigan GC-MS double stage quadrupole mass spectrometer, under the following parameters: electron energy 70 eV; electron emission 300 μA, ion source at 200 ^{o}C and transfer line at 250 ^{o}C.

Materials: In method 1, aliquot solutions containing increasing amounts of halothane (H) in the ranges 0-3 μL and 0-5 μL, in 1 mL of toluene were prepared, with a constant quantity of 1 μL dichloromethane (DCM), added as internal standard (IS) [62-64]. The sampling glass tubes (100 mm x 5 mm id) were filled with 400 mg active charcoal and mineral fibers [62-64]. Toluene, the solvent, eluted long after H and DCM, in both cases.

Sampling and Desorption: In method 1, for calibration purposes, each aliquot solution of H with 1 μL DCM was adsorbed twice for five minutes, in the charcoal filled glass tubes by using a portable membrane pump at a rate of 150 mL/min [63]. In method 2, the calibration was made by 5‘ sampling 1 μL DCM together with increasing amounts of H in the range 0-5 μL, directly onto the charcoal adsorbent (recovery of halothane 67%), at 200 mL/min. Prior to onsite sampling, 1 μL of DCM was adsorbed in the charcoal cartridges. The analyte and DCM were washed out of the charcoal in 1 mL toluene by agitating (1’). One μL of the supernatant was then injected twice into the GC.

Methods: The ratio of the H/DCM peak heights (method 1) and that of the peak areas (method 2) were determined twice for each sample. The dependence of these ratios on the amounts of halothane adsorbed/desorbed was plotted in the ranges 0-1 μL and 0-3 μL (method 1), and within 0-5 μL (method 2), respectively.

Results and Discussion

Figure 4.1 and Figure 4.2 presents the total ion chromatogram (TIC) and the SIM-GC/MS spectrum of the aliquot sample of 0.2 μL H. The specificity and the accuracy of the quantitative analysis are improved by the SIM-GC/MS technique, though the separation of the DCM and H is better in method 1 (1'30" and 1'41") than in method 2 (0.88' and 0.95'), and are closer for the shorter capillary Rtx column.

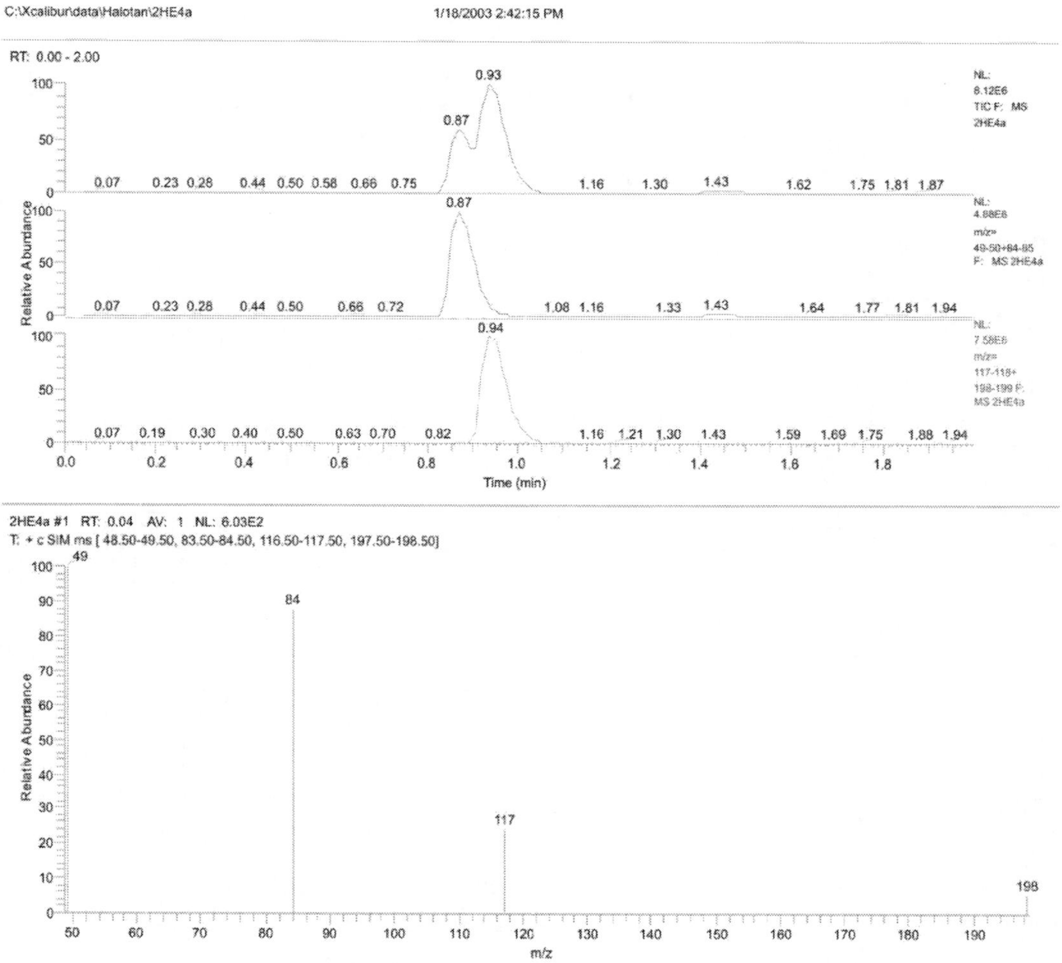

Figure 4.1 and 4.2. Total ion current (TIC) and SIM-GC/MS spectrum of 0.2μg H.

The molecular ion and a fragment ion were selected from the mass spectrum of each compound for quantitative analyses in the SIM mode, at m/z 198 and 117 (H) and at m/z 84 and 49 (DCM). The use of a suitable internal standard, in both procedures, significantly improves the accuracy of quantitative analyses [4]. The *linearity* of each method was tested in different ranges, according to the halothane concentrations found in the surgery halls:

between 0 – 1 μL H, and 0 – 3 μL H (GC method 1) or between 0 – 5 μL H (GC/MS method 2), respectively, as shown in 4.1, with good correlation coefficients of 0.99 [62-64].

For precision (RSD < 13.5%) and accuracy (RSD < 15.5%), the analytical procedures were repeated for n times for certain samples, as presented in Table 4.2. The limit of detection LOD, for a signal-to-noise ratio of 10:1, was 2 mg H/m^3 air, in both methods.

Table 4.1. Methods linearity: regression curves equations in different ranges of H concentration

Range of halothane amount / method	Equation of the regression curve	Correlation coefficient, r
0 - 3 μl (GC method 1)	y = 1.7688 x + 0.2276	0.99
0 - 1 μl (GC method 1)	y = 2.1508 x + 0.0756	0.99
0 - 5 μl (GC-MS method 2)	y = 0.8930 x + 0.0927	0.99

Table 4.2. Experimental data on the precision and accuracy of the methods, assessed for the halothane aliquot solutions of 0.2 μl, 1 μl and 2 μl, respectively. n = number of repeated experiments

Halothane aliquot solutions/method	n	Precision R. S. D. (%)	Accuracy R. S. D. (%)
0.2 μL /GC method 1	8	13.36	15.63
1 μL/GC-MS method 2	7	10.32	8.68
2 μL/GC-MS method 2	7	13.64	12.24

Table 4.3. Halothane fumes released during single general anaesthesia, in different surgery rooms. * = sample collected during two surgery operations

Crt. Nr.	Sample/method	mg H/m^3 air
1	S 1- orthopedic surgery room/GC	26
2	S 2- orthopedic surgery room/GC	18
3	S 3- general surgery room /GC	275
4	S 4 - general surgery room /GC	170
5	S 5 -general surgery room 3/GC-MS	374
6	S 6- general surgery room 4/ GC-MS	80
7	S 7* - general surgery room 5 / GC-MS	619
8	S 8 - general surgery room 6 / GC-MS	1331

The two methods described were applied in determining the halothane concentration in the air of different surgery rooms, at different hospitals. Table 4.3 shows the experimental results obtained. The air samples were collected during one single surgery operation under general anaesthesia, in each case. The sampling tubes were placed close to the breathing region of the medical staff. Seven of the eight samples were taken during one single narcosis operation. Except for the orthopaedic surgery ward, the halothane concentrations found exceeded the TLV recommended in different countries [7-9, 12, 13]. Periodic monitoring of anaesthetic gas concentrations is therefore recommended to assess the occupational exposure of medical staff.

Conclusion

The analytical quantitative procedures presented are relatively simple and rapid, the increased accuracy being provided by the use of an internal standard in both cases. The SIM-GC-MS procedure is recommended for the quantitative analysis of halothane fumes in air, due to its higher sensitivity and specificity in the SIM mode, compared to the GC - FID detection.

5. [^{15}N]-Glycineabsorption on Chronic Entheropathy and Hepatic Diseases

Introduction

Stable isotopes provide clinically useful tools for a variety of metabolic studies, offering significant advantages in obtaining quantitative information in many diseases. Administration and recovery of the labeled form of an amino acid can be combined with analysis to obtain the kinetic behavior of this nutrient. The use of tracers provides a nonradioactive and noninvasive procedure in studies of pediatric nutrition and gastroenterology.

Mass spectrometry (MS) and gas chromatography coupled with mass spectrometry (GC/MS) are the techniques used for quantitative analysis of compounds labeled with stable isotopes. These physical techniques used in medicine and other fields of science have been very much improved in the last decades. The use of the ^{15}N stable isotopein pharmacokinetic studies [71-74] and nitrogen turnover studies has frequently been reported [75-82]. The most used labeled compound is the [^{15}N]-glycine due to its lower cost.

[^{15}N]-Glycine absorption in serum and the cumulative [^{15}N]-glycine excretion in urine and faeces after a single oral administration of [^{15}N]-glycine to healthy and patients (malabsorption or children with hepatic insufficiency) were studied. Isotope analysis was performed either on nitrogen-gas from total nitrogen or directly on glycine using the GC/MS method.

Glycine plays several roles in human metabolism, e.g., as a 1-carbon donor, in purine synthesis, and as a component of glutathione. Glycine is decarboxylated via the glycine cleavage system that yields concurrent generation of a 1-carbonunit as 5, 10-methylenetetrahydrofolate. Serine hydroxymethyltransferase catalyzes the interconversion of glycine and serine, another 1-carbon donor. The quantitative role of glycine in human 1-carbon metabolism has received little attention [77, 81].

The aim of this study was to estimate malabsorption status or hepatic disturbances in humans using a stable isotope tracer. [^{15}N]-glycine, 98.98 atom %, was synthesized at the National Institute for Research and Development for Isotopic and Molecular Technology Cluj-Napoca and was administered orally as a single bolus dose to twelve patients. Six healthy children and six suspected of having malabsorption were studied. An isotope dilution GC/MS method and Kjeldahl digestion followed by MS analysis of nitrogen gas was performed. Gas chromatographic separation was performed on a glass column 2m x 3mm i.d. packed with EGA 1% on Chromosorb W AW 80-100 mesh. Validation of the parameters showed linearity in the range of 20 - 80 μg/ml and precision and accuracy with relative standard deviations better than 3%. Significant differences in the amount of [^{15}N]-Gly eliminated in urine and faeces between patients and control were obtained. More emphasis should also be given to [^{15}N]-glycine analysis by GC/MS of faeces than to ^{15}N analysis by Kjeldahl total nitrogen, to avoid the influence of nitrogen absorption by intestinal bacteria.

The distribution of ^{15}N and [^{15}N]-glycine between body fluids of infants in healthy and patients was studied, with the final intention of clinical characterization of the disease status. The investigation was extended to six children hospitalized for diseases other than digestive, called "healthy subjects," and six children suffering of severe dystrophy and enteropathy syndrome, being suspected of intestinal malabsorption and/or hepatic disturbances.

Experimental

The subjects of the study were six children with a severe dystrophy and enteropathy syndrome who were suspected of having intestinal malabsorption. Six children who were hospitalized for reasons unrelated to digestive disease were enrolled as healthy control subjects. All subjects were hospitalized in Pediatric Clinic II, Cluj-Napoca, Romania. The protocols were approved by the University of Medicine and Pharmacy from Cluj-Napoca and written informed consents were obtained from each subject parent(s) prior to this study.

Standard amino acids and trifluoroacetic anhydride were obtained from Merck (Darmstadt, Germany). Acetyl chloride and the ion exchange resin Dowex 50W-X8 were from Fluka (Buchs, Switzerland), and Ethylene Glycol Adipate was purchased from Chrompack (Middleburg, The Netherlands). [^{15}N]-glycine (Gly: 98.98%) was produced by chemical synthesis.

The analytical method has been published [74].

The single dose and the end product method were used. A single oral dose of [^{15}N]-glycine was given in the morning before breakfast followed by 24 hours urine and faeces collection for three days (with two exceptions when collection was performed for two days only). Blood samples were taken before the administration of tracer and 30, 60 and 120 minutes (or only 30 and 60 minutes in two cases) after glycine administration.

Faeces preparation: Prior to analysis, each faecal sample was homogenized (with water if needed). The total weight before and after addition of water was recorded. The faecal samples were dried at 100-105°C to constant weight, and dry weight was recorded. The dried sample was ground into a fine powder and thoroughly mixed to ensure homogeneity, before combustion for GC/MS analysis. For glycine analysis, the samples were mixed with distilled water for three minutes and subjected to deproteinization, addition of standard (20 μg ^{15}N), purification on ion exchange resin, derivatization and analysis, as described before [74]. The samples were divided into two parts and analyzed, one with internal standard addition and the other without internal standard addition. Isotope analysis was performed on the fragment ions of glycine m/z 154 and m/z 155 or on m/z 172 and m/z 173 when interferences from other compounds were observed on the glycine peak at m/z 154 and m/z 155. A resolution of about 3,000 was used to avoid interference with other compounds. Glycine concentrations were determined by isotope dilution mass spectrometry [73].

For total nitrogen analysis in urine and faeces, the Kjeldahl method was used, and isotope analysis was finally performed by isotope ratio mass spectrometry on nitrogen gas.

Calculation was made from data obtained by duplicate sample analyses, one without internal standard and one with internal standard, by using the regression curve method and the formula:

$$p = S/V\ (Cs - C)/\ (C - Cp)$$

where p is the trace quantity, S is the internal standard quantity, V is the sample volume, Cp is the abundance (atomic fraction) of tracer [^{15}N]- glycine, Cs is 99% of [^{15}N]-glycine (used as internal standard), C is the abundance (atomic fraction) of tracer in the sample with the same internal standard added.

Relative data (% of dose) rather than absolute values have been chosen to account for the balance of ^{15}N. Serum absorption of [^{15}N]-glycine was calculated assuming a mean blood content of 75 ml/kg body weight.

Results and Discussion

Clinical characteristics of the subjects, their diagnoses, and the ^{15}N doses administered are shown in Table 5.1.

In almost all cases, the maximum absorption of [^{15}N]-glycine, as reflected in serum, was already attained by 30 minutes after intake, and it did not exceed 7% of dose. Only a small fraction of the ^{15}N dose was eliminated as unmetabolized [^{15}N]-glycine, the elimination in urine ranging from 0.11 to 1.15% of the dose, while the quantity eliminated in faeces was about ten times smaller. Most of the ^{15}N was eliminated as urine total nitrogen, up to 35.8%. The total amount of ^{15}N eliminated during three days as total nitrogen in urine + faeces showed recoveries of ^{15}N between 15 and 40%. In one case of a suspect malabsorption, the elimination of ^{15}N in faeces far exceeded the total ^{15}N elimination in urine. The same subject showed also the smallest elimination of glycine in urine.

Table 5.1. Personal data of the investigated subjects

Subj.	Age	Weight (kg)	Dose mg [^{15}N]-Gly/kg	Diagnosis
HC	1y 4m	8.5	19.65	healthy
BA	14y	60	7.07	"
PB	10.5m	7.8	19.94	malabs.susp.
TV	11m	5.6	19.00	"
SN	1y 1m	6	18.28	"
CR	5y	33	9.24	healthy
CC	2y 5m	12.5	10.38	"
BI	9m	8.4	15.52	"
TT	5y	21	9.79	"
LN	17y	34	8.46	coeliac disease
BD	9m	5.4	12.15	enteropathy
CN	5m	5.8	13.15	enterocolitis

y=year; m=months; malabs. susp.=malabsorption suspects.

Table 5.2. Elimination (absorption) of [^{15}N]-Gly and ^{15}N in biological samples (% dose). Student t test of data

Serum (adsorption)					Urine			Faeces		
	% dose30'	p	[^{15}N]-Gly	p	^{15}N	P	[^{15}N]-Gly	p	^{15}N	p
C	3.82±1.89		0.53±0.05		20.7±11.5		0.046±0.02		2.37±1.70	
N	6		5		6		4		6	
PB	2.17	NS *	0.14	0.0005	12	NS	-		3.1	NS
TV	1.9	NS	0.11	0.0005	4.8	NS	-		17	0.0005
SM	2.4	NS	0.21	0.0005	16	NS	-		5.1	NS
LN	5.67	NS	0.86	0.0005	17.3	NS	0.073	NS	2.1	NS
BD	1.07	NS	1.77	>0.0005	16.5	NS	0.027	NS	2.9	NS
CN	1.25	NS	1.15	>0.0005	20.9	NS	0.182	0.005	4.1	NS

NS =not significant; n = number of control cases.

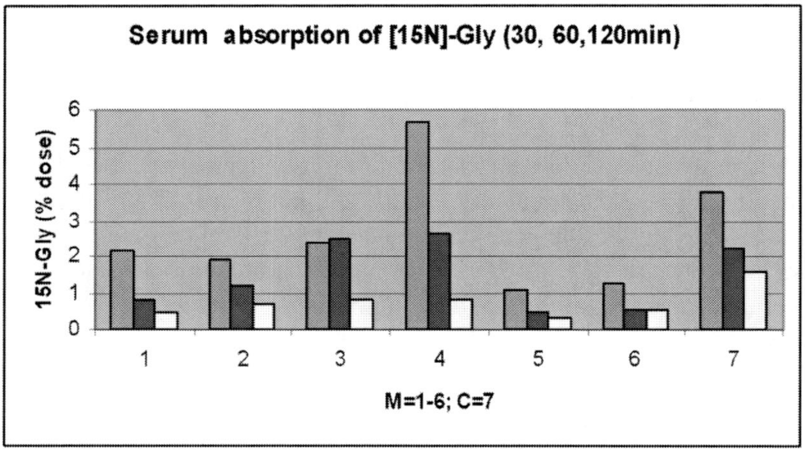

Figure 5.1. Serum absorption of [^{15}N]-glycine 30, 60 and 120 minutes after intake of the dose (1-6). The mean values are given for control group (7).

The absorption of [^{15}N]-glycine in serum at 30, 60 and 120 minutes after intake of the dose is shown in Figure 5.1, as mean values for the control group and as individual data for the patients. Figure 5.1 shows lower values in the patients if groups of ages are compared, but the differences are not significant.

Figure 5.2 shows the elimination of [^{15}N]-glycine and ^{15}N in urine, (% of dose), taking again mean values for the control group and individual data for the patients group. The Student-t-test shows significant differences between control and patients in the case of [^{15}N]-glycine.

It is to be noted that in the urines from the malabsorption group, lower elimination of [^{15}N]-glycine was observed as compared to the control group and higher values in the case of patients with hepatic disturbances. In the case of total nitrogen, the differences between patients and control subjects are not significant because of the large standard deviation obtained in the control subjects. In the case of comparison of younger subjects only, the errors could be reduced.

Figure 5.3 shows the comparison of data obtained for the faecal samples; the significance of the differences could be improved if we eliminate the constipated control subject.

The analysis of [^{15}N]-glycine in faeces was started during the investigation of the subjects with hepatic disturbances. Significantly high elimination of [^{15}N]-glycine was found in patient CN and of total ^{15}N in patient TV, which also shows the smallest total ^{15}N elimination in urine. The patients suspected of hepatic disturbances for which analysis of [^{15}N]-glycine elimination in faeces was performed present high values for ^{15}N eliminated in urine. Information on the possible influence of nitrogen absorption due to intestinal bacteria may be obtained from the comparison of total ^{15}N elimination in faeces (which includes the absorption by intestinal bacteria), with the [^{15}N]-glycine elimination, which passed as well the genuine intestinal absorption as the absorption by bacteria.

The mean values of the controls show on a percent basis about 50 times higher elimination of total ^{15}N in faeces than the elimination of [^{15}N]-glycine. On the other hand, it seems to be a fair correlation between [^{15}N]-glycine and total ^{15}N elimination in faeces (r=0.500), indicating intestinal absorption and absorption by bacteria to be consecutive processes rather than competitive processes (Figure 5.4).

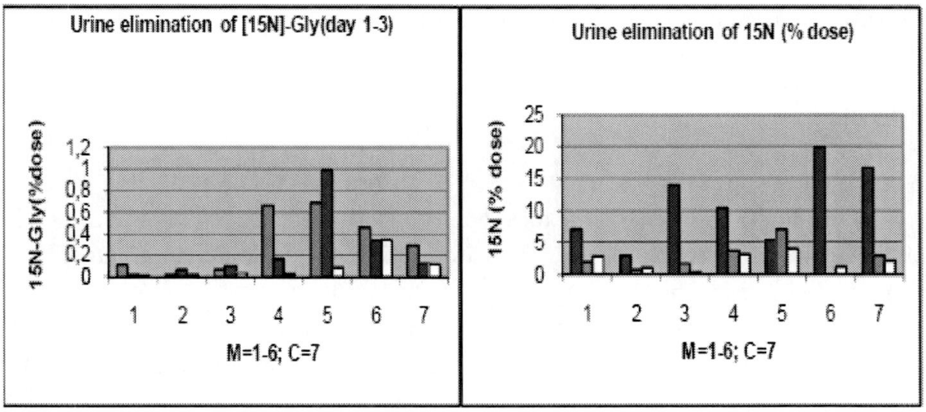

Figure 5.2. Elimination of [^{15}N]-glycine and 15N in urine, % of dose (M=1-6). The mean values for the control group (C=7) are compared with individual data of patients group.

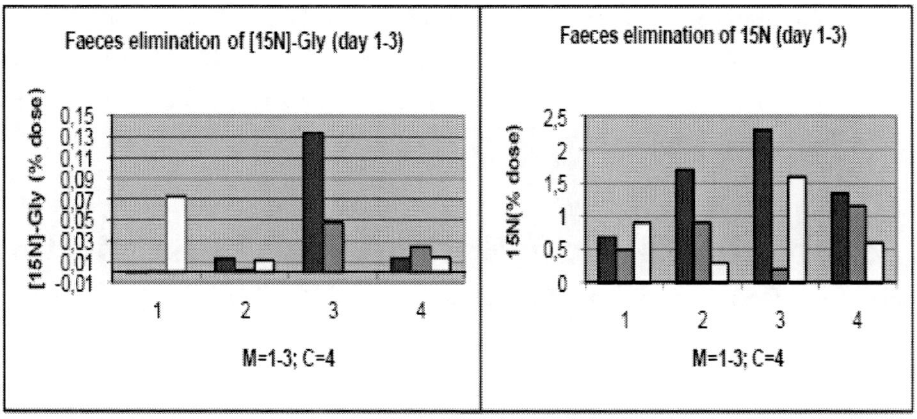

Figure 5.3. Elimination of [^{15}N]-glycine and ^{15}N elimination in faeces, % of dose (M=1-3). The mean values for the control group (C=4) are compared with individual data of patients group.

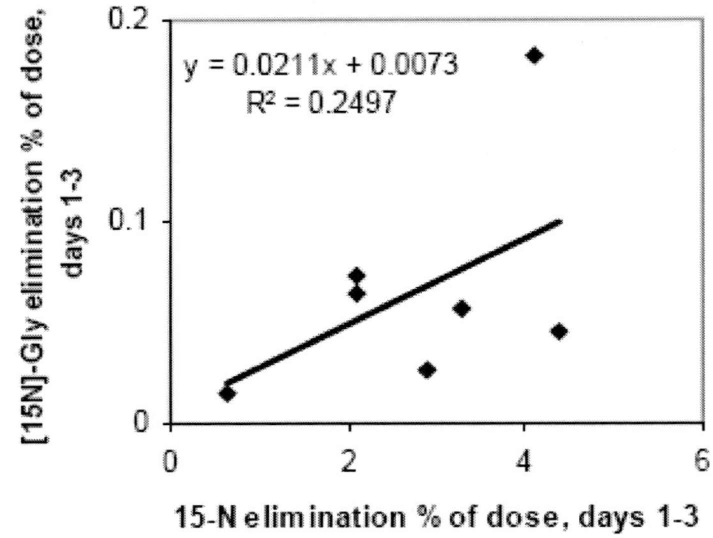

Figure 5.4. Comparison between [^{15}N]-glycine and ^{15}N elimination in faeces. The mean values for the control group are compared with individual data of patients group.

Conclusion

No statistically significant differences for the absorption of [^{15}N]-glycine, as reflected in serum, were obtained between the healthy controls and patients. No significant differences were obtained for the elimination of ^{15}N in total nitrogen of urine. In change, significant differences, in both directions, were observed for the elimination of [^{15}N]-glycine in urine of patients as compared to healthy controls. Significantly increased elimination of ^{15}N in faeces was observed for total ^{15}N in one of three malabsorption children and for [^{15}N]-glycine in one of three hepatic disturbances suspects investigated, the other two having the percentiles 9 and 44. More emphasis should be given to faeces data than to urine because ^{15}N elimination in urine is competitive with ^{15}N incorporation to proteins. More emphasis should also be given to [^{15}N]-glycine analysis by GC/MS of faeces than to ^{15}N analysis by Kjeldahl total nitrogen, to avoid the influence of nitrogen absorption by intestinal bacteria.

The poor statistical interferences could be due to different ages of the control subjects and to the fact that the controls, although digestively healthy, were in fact children hospitalized for other diseases.

6. Isotopic Dilution Methods for Theophylline Determination in Plasma and Saliva of infants

Introduction

Theophylline was a frequently used drug in the neonatal intensive care units for infants with apnea of prematurity. In children with asthma, slow-release theophylline is a common choice to maintain a good control of the disease. Multiple factors such as age, some drugs (e.g., erythromycin, cimetidine), hepatic or kidney dysfunction, fever or food, may influence theophylline pharmacokinetics in a given patient. The therapeutic level of the drug in plasma is between 5-15 μg/ml. The narrow therapeutic index, the large interpatient variability in dosage requirements due to variability of theophylline pharmacokinetics and the poor correlation between dosage and plasma levels have increased the need of individualized dosing of this drug. Salivary levels of theophylline as an alternative for plasmatic levels theophylline treatment monitoring showed a ratio plasma/saliva that ranged between 1.87±0.05 to 1.45±0.02 [83-86]. Good correlation was reported (r=0.99) for two methods of measuring theophylline, an immuno-enzymatique and a gas chromatographic one [87]. Theophylline is protein-bonded in serum 52-65% [88] and has a circardian rhythm [85]. Studies on theophylline pharmacokinetics in infants showed that postconceptional age rather than postnatal age should be used as a maturation marker during theophylline therapy [89]. The aim of this study is to evaluate the efficiency of salivary theophylline determination by using two isotope dilution mass spectrometric techniques to monitor the treatment with theophylline in infants.

Experimental

^{15}N-theophylline, 74.2 atom % ^{15}N, labeled at the nitrogen in the position seven, synthesized at the Institute of Isotopic and Molecular Technology Cluj-Napoca, Romania, was used as internal standard [90]. The purity was ascertained by infrared spectroscopy, mass spectrometry and melting point. Theophylline was administered orally in children with asthma using Teo-Dur (Key Pharmaceuticals, Inc., USA), while infants received I.V. aminophylline. A MAT 311 EI mass spectrometer was used in the following conditions: resolution 3,000, electron energy 70eV, electron emission 100μA, ion source 200°C. The direct inlet probe was used in the automatic heating mode from 20-250°C in 200s. Comparatively, a Hewlett Packard 5989B mass spectrometer coupled to a 5890 gas chromatograph was used in the conditions: electron energy 70 eV; electron emission 300μA and ion source temperature 200°C.

The gas chromatograph-mass spectrometer (GC/MS) assay used a HP-5MS fused silica capillary column, 30mx0.25mm, 0.25μm film-thickness, programmed from 200°C to 270°C at 10°C/min, the flow rate 1ml/min, with helium as carrier gas. Retention time for theophylline and internal standard was 4.2min. The ion m/z 180 for theophylline and the molecular ion m/z 181 for the internal standard were monitored for quantitative analyses in the selected ion monitoring (SIM) mode.

Population. Theophylline concentration measurements were performed in 40 hospitalized infants and children treated with theophylline for apnea of prematurity or asthma. Two different groups were studied: A, formed by 27 children with asthma aged 2-16 years old, treated with slow-released theophylline and B, consisting of 13 infants with apnea of prematurity, aged between 2-10 weeks. The main dose was 15.1mg/kg/24h, for group A. Blood and saliva-paired samples were taken, at four hours and at eight hours after morning dose. In group B I.V., aminophylline was administered in a loading dose of 5mg/kg and maintenance doses of 3mg/kg, every eight hours. Simultaneous plasma and saliva, at two and four hours after loading dose and at four and eight hours in the second and third day were sampled. Written informed consents were obtained from each subject's parent prior to this study.

Calculation: Regression curves obtain by two different mass spectrometric methods and the least squares with matrix calculation were compared to correlate the levels of the drug.

Table 6.1. Comparative values of plasma and saliva levels of theophylline in group A and group B

Population	Range, μg/ml	Mean ± SD, μg/ml
Plasma levels		
Group A=Children	1.98 – 21.96	7.98 ± 5.25
Group B=Infants	1.62 – 27.90	7.76 ± 5.85
Saliva levels		
Group A=Children	1.41 – 15.06	5.12 ± 3.45
Group B=Infants	1.02 – 18.23	5.51 ± 4.61
Plasma / saliva ratio		
Group A=Children	1.29 – 2.05	1.63 ± 0.23
Group B=Infants	1.14 – 2.01	1.53 ± 0.28

Results and Discussion

Direct inlet mass spectrometric assay was validated in the range 0-40μg/ml of theophylline [90]. The regression curve used for calculation, obtained by using standard samples, was y=0.103x+0.366.

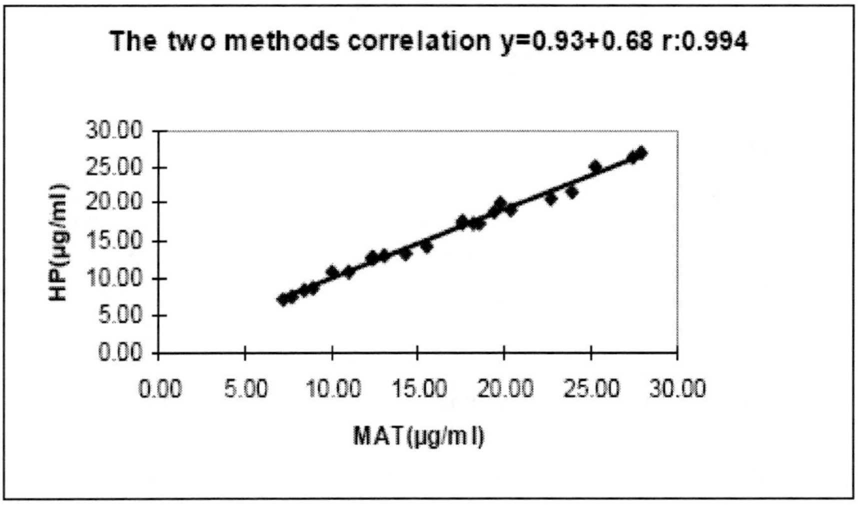

Figure 6.1. Correlation between the two methods for theophylline determination.

In the gas chromatograph-mass spectrometric (GC/MS) assay, m/z 180 and 181 ions were measured in the selected ion monitoring (SIM) mode, for theophylline and the internal standard, respectively. The regression curve obtained in GC/MS assay was y=0.12x +0.38, r=0.990. Precision gave R.S.D values lower than 8% for 10μg/ml (n=8) and accuracy lower than 6%. Sensitivity measured at a signal/noise 4/1 was 0.5μg/ml. The same concentrations were also obtained by matrix calculation according to the least squares method (r=0.999).

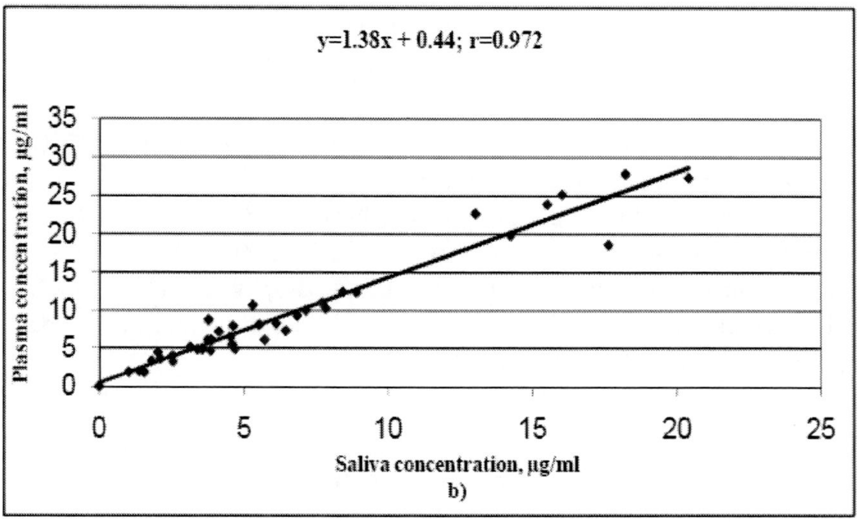

Figure 6.2. Correlation between plasma and saliva theophylline in infants.

Each value was obtained as an average between two measurements of the same sample. Correlation curves between the two methods presented in Figure1 gave for saliva r=0.998 and for plasma r=0.993. The correlation curve between the two different methods of calculation was good (r=0.998). Table 6.1 shows the ratios values and correlation curves between plasma and saliva obtained for the two study groups: 1.63±0.23 (R.S.D:6.9%) for group A and 1.53±0.28 (R.S.D. 5.4%) for group B. The good correlation (r=0.97) between plasma and saliva for the two groups is represented in Figure 6.2.

The results confirm that the measurements of drug concentration levels in saliva provide adequate reliability for therapeutic drug monitoring of theophylline in asthma and in apnea of prematurity. Our data revealed lower ratio between plasma and saliva concentration in infants than in children. The difference reflects the "physiological" hypoalbuminemia in premature infants due to the fact that salivary theophylline is concordant with unbounded plasma theophylline.

Conclusion

The isotope dilution mass spectrometric methods used are simple, precise and rapid.

The regression curve method of calculation gave similar results as the matrix calculation method (r=0.998).

A very good correlation was obtained between two methods, using a double focusing mass spectrometer in high-resolution mode and a quadrupole one coupled to a gas chromatograph (GC/MS).

Our results confirm that the measurements of drug concentration levels in saliva provide adequate reliability for therapeutic drug monitoring of theophylline in asthma and in apnea of prematurity.

Acknowledgments

Versions of this paper have been published in Romanian J Biophys, 2007 17(3):169 and 2010 20(1): 7.(Subchapter 1), in Studia Physica Cluj-Napoca, 2001, 385-389 and Asian J Chem, 2010, 22,3608 (Subchapter 2) in "Chemické listy" Journal, 2008,102,s963 (Subchapter 3), in Indoor and Bult Environment, 12 (2003) 125 and 14 (2005) 405 by SAGE Publications Ltd, All rights reserved (subchapter 4), in Romanian J. Biophysics, 4 (1994) 43 (subchapter 5), in Chromatographia, 53 (2001) S387, with kind permission from Springer Science+Business Media B.V.(subchapter 6).

References

[1] W. B. Dunn, D.I. Ellis, *Trends Anal. Chem.*, 24 (2005) 285.
[2] S.R. Thorpe, J.W. Baynes, *Amino Acids*, 25 (2002) 275.
[3] M. Culea, D.L. Hachey, *Rapid Commun. Mass Spectrom.* 9 (1995) 955.
[4] C. Deng, Y. Deng, *J. Chromatogr. B Analyt. Technol. Biomed. Life Sci.*, 792 (2003) 261.
[5] O. Calatayud, M. Rodriguez, A. Sánchez-Alcrazt, P. Ibánez, *J. Clin. Pharm. Therap.*, 20 (2008) 23.
[6] T. Flatt, *Nature*, 462 (2009) 989.
[7] R.C. Grandison, M.D. Piper, L. Partridge, *Nature*, 462 (2009) 1061.
[8] X. Gao, *Science,* 324 (2009) 1565.
[9] J. Fan, M. Sammalkorpi, M. Haataja *Phys. Rev.* E 81, (2010) 011908.
[10] A. Arias, A. Ormazabal, J. Moreno, B. González, M.A. Vilaseca, J. Garcíavilloria, T. Pàmpols, P. Briones, R. Artuch, A. Ribes, *J. Neurosci. Methods*, 156 (2006) 305.
[11] S.M. Carter, J.M. Midgley, D.G. Watson, R.W. Logan, *J. Pharm. Biomed. Anal.,* 9 (1991) 969.
[12] D.H. Chace, J.E. Sherwin, S.L. Hillman, F. Lorey, G.C. Cunningham, *Clin. Chem.* 44 (1998) 2405.
[13] D.H. Chace, *Chem. Rev.*, 101 (2001) 445.
[14] M.L. Couce Pico, D.E. Castiñeiras Ramos, M. López Sousa, M.J. Fernández Seara, Eirís, J. Puñal, J.A. Cocho De Juan, *An. Pediatr. (Barc.)*, 69 (2008) 239.
[15] M. Culea, A. Iordache, C. Mesaros, *Chemické listy Journal*, 102 (2008) s636.
[16] M. Culea, M. Chiriac, E. Culea, C. Lehene, A. Iordache, *Romanian J. Biophys.*, 17 (2007) 169.
[17] C. Deng, C. Shang, Y. Hu, X. Zhang, *J. Chromatogr. B*, 775 (2002) 115.
[18] C Deng, X. Yin, L. Zhang, X. Zhang, *Rapid Commun. Mass Spectrom.*, 19 (2005) 2227.
[19] O. Fiehn, T. Kind, *Methods Mol. Biol.*, 358 (2007) 3.
[20] D.T. Forman, *Ann. Clin. Lab. Sci.*, 21 [1991] 85.

[21] E. Gelpi, *J. Chromatography A,* 703 (1995) 59.

[22] T. Hagen, M.S. Korson, M. Sakamoto, J.E. Evans, *Clin. Chim. Acta,* 283 (1999) 77.

[23] W.C. Heird, R.B. Dell, R.A. Helms, H.L. Greene, M.E. Ament, P. Karna, M.C. Storm, *Pediatrics*, 80 (1987) 401.

[24] J.W. Hou, 46 (2005) 59.

[25] Y. Inoue, M. Ohse, T. Shinka, T. Kuhara, *J. Chromatogr. B, Analyt. Technol. Biomed. Life Sci.*, 870(2008)160.

[26] P. Jouvet, P. Hubert, J.M. Saudubray, D. Rabier, N.K. Man, *Pediatr. Res.*,58(2005) 278.

[27] W.D. Lin, J.Y. Wu, C.C. Lai, F.J. Tsai, C.H. Tsai, S.P. Lin, D.M. Niu, *Acta. Paediatr. Taiwan,* 42(2001) 224.

[28] D.S. Millington, N. Kodo, N. Terada, D. Roe, D.H. Chace, *Intl. J. Mass Spectrom. Ion Process*, 111 (1991) 211.

[29] K. Murayama, M. Kimura, S. Yamaguchi, T. Shinka, K. Kodama, *Brain Dev.*, 19 (1997) 303.

[30] K. Nagy, Z. Takats, F. Pollreisz, T. Szabo, K. Vekey, *Rapid Commun. Mass Spectrom.*, 17 (2003) 983.

[31] A. Niederwieser, A. Matasovic, P. Tippett, D.M. Danks, *Clin. Chim. Acta,* 76 (1977) 345.

[32] K.K. Pasikanti, P.C. Ho, E.C. Chan, *J. Chromatogr. B. Analyt. Technol. Biomed. Life Sci.* 871 (2008) 202.

[33] Y-Z. Song, B-X. Li, H. Hao, R.-L. Xin, T. Zhang, C.-H. Zhang, K. Kobayashi, Z.-N. Wang, X.-Y. Zheng, *Clinical Biochemistry*, 41(2008) 616.

[34] E.J. Want, B.F. Cravatt, G. Siuzdak, *Chembiochem.*, 6(2005) 1941.

[35] B. Wilcken, Ch.B., V. Wiley, J. Hammond, K. Carpenter, *N. Engl. J. Med.*, 348 (2003) 2304.

[36] R.C. Willis, *Modern Drug Discovery*, 5 (2002) 28, Http://Pubs.Acs.Org/ Subscribe/ Journals/Mdd/V05/I04/Html/04willis.Html.

[37] Http://Www.Lifespan.Org/Adam/Healthillustratedencyclopedia/1/003361.Html.

[38] M. Culea, N. Palibroda, V. Mercea, A.D. Abraham, *Rev. Roum. Biochim.*, 23 (1986) 273.

[39] S. Nicoara, M. Culea, A. Nica, E. Culea and O. Cozar, *Indoor Built Environ.*, 14 (2005) 405.

[40] M. Culea, N. Palibroda and A.D. Abraham, *Rev. Roum. Biochim.*, 24 (1987) 169.

[41] J. Jodynis-Liebert, J. Flieger, A. Matuszewska, J. Juszczyk , *J. Clin. Pharmacol.*, 44 (2004) 338.

[42] M. Culea and D.L. Hachey, *Rapid Commun. Mass Spectrom.*, 9 (1995) 655.

[43] M. Culea, N. Palibroda, P. Panta Chereches and M. Nanulescu, *Chromatographia*, 53 (2001) S387.

[44] A.P. Ferreira, *Cad. Saude Publica Rio de Janeiro*, 21 (2005) 1884.

[45] G. Host, A. Wahllander, U. Von Mandach, R. Preisig, *Hepatology*, 7 (1987) 338.

[46] N.R. Scott, D. Stambuk, J. Chakraborty, V. Marks, M.Y. Morgan, *Clin. Sci.*, 74 (1988) 377.

[47] A. Hostege, M. Staiger, K. Haag, W. Gerok, *Klin. Wochenschr.*, 67 (1989) 6.

[48] W.F. Balistreri, H.H. A-Kader, K.D, Setchell, D. Gremse, F.C. Ryckman, T.J. Schroeder, *Ann. Clin. Lab. Sci.*, 22 (1992) 162.

[49] P. Tangkijvanich, V. Mahachai, S. Wittayalertpanya, V. Ariyawongsopon, S. Isarasena, *Southeast Asian J. Trop Med. Public Health*, 31 (2000) 152.
[50] Y.C. Bechtel, H. Lelouet, M.P. Brientini, M. David-Laroche, J.P. Miguet, G. Paintaud, P.R. Bechtel, *Therapie*, 55 (2000) 619.
[51] M.C. Bastien, F. Leblond, V. Pichette, J.P. Villeneuve, *Can. J. Physiol. Pharmacol.*, 78 (2000) 912.
[52] J.L. Brazier, B. Ribon, M. Desage, B. Salle, *Biomed. Mass Spectrom.*, 7 (1980) 189.
[53] S. Wittayalertpanya, V. Mahachai, *J. Med. Assoc. Thai*, 84, (2001) S189.
[54] Y. Benchekroun, M. Desage, B. Ribon, J.L. Brazier, *J. Chromatogr.*, 532 (1990) 261.
[55] M. A. Shirley, R. C. Murphy, *J. Biol. Chem,* 265 (1990) 16288.
[56] P. Cap, J. Chladek, F. Pejhal, M. Maly, V. Petru, P.J. Barnes, P. Montuschi, *Thorax*, 59 (2004) 265.
[57] B. Prokes, *Med. Pregl.* 51 (1998) 532.
[58] S. Sardas, N. Aygun, M. Gamli, Y. Unal, N. Unal, N. Berk, A.E. Karakaya, *Mutat. Res.* 418(1998) 93.
[59] C.J. Cote, *Anesthesiology,* 94 (2001) 934.
[60] J.P. Morray, J.M. Geiduschek, C. Ramamoorthy, C. M. Haberkern, A. Hackel, R.A. Caplan, K.B. Domino, K. Posner, F.W. Cheney, *Anesthesiology* 93 (2000) 6.
[61] K.A. Henderson, I.P. Matthews, *Eur. J. Anaesthesiology* 94 (2001) 149.
[62] S. Nicoara, M. Culea, N. Palibroda, O. Cozar, *Indoor Environ.*3 (1994) 83.
[63] M. Culea, S. Nicoara, A.N. Nica, C. Carmen, *Indoor Built Environm.* 12 (2003) 125.
[64] S. Nicoara, M. Culea, A. N. Nica, E. Culea, O. Cozar, *Indoor Built Environm.* 14 (2005) 405.
[65] J.P. Barker *Anaesthesia* 52 (1997) 1077.
[66] G. Bozkurt, D. Memis, G. Karabogaz, Z. Pamukcu, M. Ture, B. Karamanlioglu B, I. Gunday, C. Algunes, *Anaesth. Intensive Care*. 30 (2002) 597.
[67] J. Friedman, *Teratology*. 73 {1988) 69.
[68] L. *Holcatova*, V. Benesova, D. Hartlova, Indoor Built Environm. *12* (2003)*121.*
[69] R. E. Korczynsky, *Applied Occup. Env. Hygiene*. 14 (1999) 384.
[70] D. Vecchio, A.J. Sasco, C. I. Cann, *Am. J. Ind. Med.* 43(2003) 369.
[71] M. Culea, N. Palibroda, Z. Moldovan, A.D. Abraham, P.T. Frangopol, *Chromatographia*, 28 (1989) 24.
[72] M. Culea, N. Palibroda, Maria Chiriac, Z. Moldovan,*Biomed. Environm. Mass Spectrom.*, 19 (1990) 360.
[73] M. Culea, N. Palibroda, Z. Moldovan, Maria Chiriac, *Rev. roum.Biochim* 28 (1991) 61.
[74] M. Culea, N. Palibroda, Maria Chiriac, Z. Moldovan, A.D. Abraham, P.T. Frangopol, *Biol. Mass Spectrom.,* 20 (1991) 740.
[75] H. Faust, K. Jung, P. Krumbiegel, et al. *ZFI-Mittelungen*, Leipzig, 1989, 150, 161.
[76] H. Faust, P. Krumbiegel, K. Jung, , *ZFI-Mittelungen*, Leipzig, 1989, 150, 151.
[77] A. Lapidot, J. Nissim, C. S. Irving, et al., *Stable isotopes: in Proceedings of the Third International Conference, Klein ER, Klein PD, New York,* (1979) 599.
[78] I. Nissim, A. Lapidot, *Biochem. Med.* 31 (1984) 185.
[79] T.P Stein, R.G. Settle, J.A. Albina et al., *J. Nutr.* 116 (1986) 1651.
[80] K. Wetzel, P. Junghans, H. Faust, B. Dudek, *Stable isotopes: in Proceedings of the Third International Conference. Klein ER, Klein PD. New York*, (1979)591.

[81] Y. Lamers, J. Williamson, L. R. Gilbert, P. W. Stacpoole, J. F. Gregory III, *J. Nutr.* 137 (2007) 2647.

[82] M. Culea, N. Palibroda, M. Chiriac, Z. Moldovan, N. Miu, P.T. Frangopol, *Romanian J. Biophysics,* 4 (1994) 43.

[83] M. Aviram, A. Tal, Z. Ben-Zvi, R. Gorodisher R , *Pediatrics*, 80(1987) 894.

[84] E. Iwasaki, M. Baba, *J. Asthma* 24 (1987)173.

[85] J.C. Jr., Virchow, J. Mansfeld, H. Mattys, C. Sr Virchow, *Pediatr. Pulmonol.* 14(1992) 37.

[86] F. Sallans, F. Rodriguez, P. Michaud, J.L. Lepargneur, *Path. Biol.* 36 (1988) 149.

[87] E. Bidat, A. Leneveu, P. Aergerter, A. Roux, B. Flouvat, J.P. Gallet *Presse Med.* 19 (1990) 158.

[88] L.A. Ferron, J-P. Weber, J. Herbert, P.M. Bedard, *Clin. Chem.* 31 (1985) 1415.

[89] D.M. Kraus, J.H. Ficher, S.J. Reitz, S.A. Kecskes, T.F. Yeh, K.M. McCulloch, E.C. Tung, M.J. Cwik, *Clin. Pharmacol. Ther.* 54 (1993) 351.

[90] M. Culea, N. Palibroda, M. Chiriac, Z. Moldovan, *Biomed. Environm. Mass Spectrom.*, 19(1990) 360.

In: Medical Physics
Editor: Marcin Balcerzyk
ISBN: 978-1-62257-590-9

Chapter II

Upscaling Microbial Motility in Porous Media

Moongyu Park*
Department of Mathematical Sciences,
University of Alabama in Huntsville, AL, US

Abstract

Drug-resistant infection is a major concern within the field of medical science. It is well known that infection is the result of bacterial biofilms that are structured aggregates in bacteria. In this chapter, we focus on the individual dynamics of a single microbe and review statistical mechanical approach within a system for such particles that have a relatively low concentration. One may think of this approach as the initial stage before the biofilms occur. Since bacterial behavior consists of a sequence of runs, a linear movement in a particular direction, followed by tumbles, random changes of direction, bacterial motility is modeled by Lévy motions. The Lévy motions are studied in three-scale porous media. Tissues, cells, folded proteins, and whole plants and animals are natural biological examples of porous media. On the micro-scale the bacterial transport is governed by a stochastic ordinary differential equation with stationary, ergodic, Markovian drift velocity and a stable Lévy diffusion. The homogenization from micro to meso and meso to macro scale is handled with generalized central limit theorems. On the mesoscale the Lagrangian drift is assumed as Lévy motions to account for the fractal character of many natural porous systems. Diffusion on the mesoscale is a result of the microscale asymptotics while diffusion on the macroscale results from the mesoscale asymptotics. Renormalized (or upscaled) Fokker-Planck equations with time-dependent dispersion tensors and fractional derivatives are presented at the macroscale. To generalize the approach to anisotropic porous media, operator-stable Lévy motions are applied to this problem.

*E-mail address: mp0002@uah.edu

PACS: 05.45-a, 52.35.Mw, 96.50.Fm

Keywords: Microbial motility, Transport in porous media, Lévy motions, Central Limit Theorems

AMS Subject Classification: 53D, 37C, 65P

1. Introduction

Infection by drug-resistant bacteria [1,2], such as Methicillin-Resistant Staphylococcus Aureus (MRSA), Streptococcus pneumoniae, and Extensively Drug-Resistant Tuberculosis (XDR TB), have been reported in many countries such as Brazil, UK, Canada, and USA. Bacteria has developed resistance to commonly used antibiotics and has transferred resistance genes to other circulating strains. Genes conferring resistance to tetracycine, erythromycin, and vancomycin have been detected and characterized in *Lactococcus actis, Enterococci* and, recently, in *Lactobacius* species isolated from fermented meat and milk products [3].

It is well known that the infection is the result of bacterial biofilms [4,5] that are structured aggregates of bacteria. In this chapter, we focus on the individual dynamics of a single microbe and review statistical mechanical approach within a system for such particles that have a relatively low concentration in porous media. One may think of this approach as the initial stage before the biofilms occur.

Throughout nature and modern technologies almost everything is porous when analyzed in a proper scale. Tissues, cells, folded proteins, and whole plants, and animals are natural biological examples of porous media; and where soils, aquifers, reservoirs, and the earth's lithosphere are geophysical examples. Modern technological examples involve chromatography, the atomic and shear forces microscopes, drug delivery substrates, ceramics, protective clothing, and water softening.

We consider porous media with fractal [6,7] functionality on the mesoscale which is assumed to lie between an upper and a lower cut off [8]. Below the lower cut off (microscale) and above the upper cut off (macroscale) the system is non-fractal. On the microscale the trajectory of a particle is assumed to be the solution of a stochastic ordinary differential equation (SODE) with stationary, ergodic, Markov drift velocity. To account for particles that may be self-motile, such as microbes [9], the microscale diffusion is assumed to be driven by an α-stable ($0 < \alpha \leq 2$) Lévy process [10]. If $\alpha = 2$, then the process is Brownian and the model can account for non-motile particles. The homogenization from micro to meso and meso to macro scale is handled with generalized central limit theorems (CLT). On the mesoscale the Lagrangian drift is assumed Lévy to account for the fractal character of many natural porous systems. Diffusion on the mesoscale is a result of the microscale asymptotics while diffusion on the macroscale results from the mesoscale asymptotics. This approach can be shown equivalent to a rigorous renormalization group approach. Renormalized (or upscaled) Fokker-Planck equations with time-dependent dispersion tensors and fractional derivatives are presented at the macroscale.

A generalized approach to anisotropic porous media operator-stable Lévy motions [11-13] are applied to this problem. The goal is to be able to write Fokker-Planck equations for each of the three scales considered. This is done by first deriving the stochastic ordinary differential equations (SODE's) on the meso- and macro-scales via the application of a generalized central limit theorems [14,15]. The Fokker-Planck equations follow directly once the SODE's are obtained on each scale. These equations are in general fractional with different fractal dimensions in the principle directions.

2. Lévy Motions

Consider the following integrated stochastic ordinary differential equation (SODE) in three dimensions:

$$\mathbf{X}(t) = \mathbf{X}(0) + \int_0^t \mathbf{V}(r)\,dr + \mathbf{L}(t), \quad t \geq 0, \tag{1}$$

where the stochastic processes $\{\mathbf{V}(r)\}$ and $\{\mathbf{L}(t)\}$ are stable Lévy processes [16,17]. Let $\tilde{\mathbf{V}}(t) = \mathbf{V}(t) - \mathbf{V}(0)$ and $\tilde{\mathbf{X}}(t) = \mathbf{X}(t) - \mathbf{X}(0)$. Then, for each t, $\tilde{\mathbf{V}}(t)$ has an α-stable distribution [16,18] ($\tilde{\mathbf{V}}(t) \sim S_{\alpha_v}(tM_v, t\mu_v)$), whose characteristic function ϕ_v is

$$\phi_v(t,\mathbf{k}) = \exp\left[-t\int_{S_3} |\mathbf{k}\cdot\mathbf{s}|^{\alpha_v}\left(1 - i\,sgn(\mathbf{k}\cdot\mathbf{s})\tan\left(\frac{\alpha_v\pi}{2}\right)\right) M_v(d\mathbf{s}) + it\mathbf{k}\cdot\mu_v)\right], \tag{2}$$

where $0 < \alpha_v \leq 2$, $\alpha_v \neq 1$, $\mathbf{k}\cdot\mathbf{s}$ is the inner product, $sgn(x) = 1$ if $x > 0$, 0 if $x = 0$ and -1 otherwise, and M is a finite measure on the unit sphere $S_3 = \{\mathbf{s} \in R^3\colon\ |\mathbf{s}| = 1\}$ [17]. When $\alpha_v = 1$

$$\phi_v(t,\mathbf{k}) = \exp\left[-t\int_{S_3} |\mathbf{k}\cdot\mathbf{s}|\left(1 + i\frac{2}{\pi}sgn(\mathbf{k}\cdot\mathbf{s})\log|\mathbf{k}\cdot\mathbf{s}|\right) M_v(d\mathbf{s}) + it\mathbf{k}\cdot\mu_v\right]. \tag{3}$$

For notational purposes, the characteristic function for $\mathbf{L}(t)$ is the same as that for $\tilde{\mathbf{V}}(t)$ with the subscript v replaced by ℓ. Let

$$\mathbf{Y}(t) = \int_0^t \mathbf{V}(r)\,dr.$$

Then $\mathbf{Y}(t)$ can be decomposed as $\mathbf{Y}(t) = \int_0^t \left(\mathbf{V}(0) + \tilde{\mathbf{V}}(r)\right)dr = t\mathbf{V}(0) + \tilde{\mathbf{Y}}(t)$, where $\tilde{\mathbf{Y}}(t) = \int_0^t \tilde{\mathbf{V}}(r)\,dr$. Therefore, we can remark (1) as

$$\tilde{\mathbf{X}}(t) = t\mathbf{V}(0) + \tilde{\mathbf{Y}}(t) + \mathbf{L}(t). \tag{4}$$

Henceforth, we assume $\alpha_v \neq 1$.

Now, $\tilde{\mathbf{V}}(t)$ is Lévy; so, it at most has a countable set of discontinuities [19] Hence, the integral in $\tilde{\mathbf{Y}}$ can be interpreted in the Riemann sense and its characteristic function may be evaluated [8]

$$\begin{aligned}\phi_{\tilde{\mathbf{Y}}}(t,\mathbf{k}) &= E\left[\exp(i\mathbf{k}\cdot\tilde{\mathbf{Y}}(t))\right] \\ &= \exp\left[-\frac{t^{1+\alpha_v}}{1+\alpha_v}\int_{S^3} |\mathbf{k}\cdot\mathbf{s}|^{\alpha_v}\psi(\alpha_v,\mathbf{k},\mathbf{s})M_v(d\mathbf{s}) + i\frac{t^2}{2}\mathbf{k}\cdot\mu_v\right],\end{aligned} \tag{5}$$

where $\psi(\alpha_v,\mathbf{k},\mathbf{s}) = 1 - i\,sgn(\mathbf{k}\cdot\mathbf{s})\tan(\alpha_v\pi/2)$. Equation (5) is quite important because it shows that in addition to being non-Markovian, $\tilde{\mathbf{Y}}$ does not even have stationary increments.

We next discuss the Fokker-Planck (Eulerian dispersion) equation that $\tilde{\mathbf{Y}}$ satisfies on the mesoscale. Let $f_{\tilde{\mathbf{Y}}}$ be the transition density for $\tilde{\mathbf{Y}} - t^2\mu_v/2$. It can be shown [8] that $f_{\tilde{\mathbf{Y}}}$ satisfies

$$\frac{\partial f_{\tilde{\mathbf{Y}}}}{\partial t} = D(t)\,\nabla^{\alpha_v} f_{\tilde{\mathbf{Y}}}, \tag{6}$$

where $D(t) = -t^{\alpha_v}/\cos(\pi\alpha_v/2) > 0$ and $\nabla^{\alpha_v} f_{\tilde{\mathbf{Y}}}$ is defined by

$$\widehat{(\nabla^{\alpha_v} f_{\tilde{\mathbf{Y}}})} = \left[\int_{S_3} |-\mathbf{k}\cdot\mathbf{s}|^{\alpha_v} \exp\left(i(-\mathbf{k}\cdot\mathbf{s})\frac{\pi\alpha_v}{2}\right) M(ds)\right] \hat{f}_{\tilde{\mathbf{Y}}}(\mathbf{k},t). \tag{7}$$

Here, $\hat{f}_{\tilde{\mathbf{Y}}}$ is the Fourier transform of $f_{\tilde{\mathbf{Y}}}$. Note that $|-\mathbf{k}\cdot\mathbf{s}|^{\alpha_v}\exp(i(-\mathbf{k}\cdot\mathbf{s})(\pi\alpha_v/2))$ is one value of the multi-valued $(-i\mathbf{k}\cdot\mathbf{s})^{\alpha_v}$ [20].

This is a fractional equation similar to the diffusion equation for Lévy processes [21] except $D(t)$ is a function of t. If $\alpha_v = 2$, then (6) is a classical anomalous diffusion equation for the evolution of the particle under convection alone. In atmospheric turbulence this result is known as Richardson turbulent diffusion [22].

A useful and well known property of α-stable Lévy motions is that their paths are fractals. There are several fractal dimensions in use such as box dimension, divider (or compass) dimension, Haussdorf dimension, etc [6,7,23]. If one takes the fractal dimension to be the box dimension, D_b then the fractal dimension of a 3-dimensional α-stable Levy process is $D_b = \max\{1, 4 - 1/\alpha\}$ [7] If one takes the fractal dimension to be the divider dimension, D_d, then $D_d = \alpha$.

In next section, we follow the classical Central Limit Theorem [24] and refer also to other works [8] for detail.

3. Scale Change and Central Limit Theorem

On the microscale, the trajectory is assumed of the form

$$\mathbf{X}^{(0)}(t) = \mathbf{X}^{(0)}(0) + \int_0^t \mathbf{v}^{(0)}(\mathbf{X}^{(0)}(r))\,dr + \mathbf{L}^{(0)}(t). \tag{8}$$

The superscript (0) refers to the microscale. We use superscripts (1) and (2) for mesoscale and macroscale, respectively. It is assumed that time, t, is positive throughout the text. Here, the drift velocity $\mathbf{v}^{(0)}(\mathbf{X}^{(0)}(r))$ is stationary, ergodic and Markovian with a mean velocity $\bar{\mathbf{v}}$ and the fluctuating position, $\mathbf{L}^{(0)}(t)$, is an α_l-stable Lévy motion that accounts for self-motility (in the case of a microbe). If $\alpha_l = 2$, then $\mathbf{L}^{(0)}(t)$ reduces to $\mathbf{B}^{(0)}(t)$, a Brownian motion, and the particle need not be motile. By using a classical central limit theorem and the fact that Lévy motions dominate Brownian motions, it can be shown for large n, $\mathbf{X}^{(0)}(nt) \approx \mathbf{L}^{(1)}(t)$, $t > 0$, where $\mathbf{L}^{(1)}(t)$ is a mesoscale Lévy motion [8,25-27]. This is the renormalization for $\mathbf{X}^{(0)}$.

The trajectory on the mesoscale is thus

$$\mathbf{X}^{(1)}(t) = \mathbf{X}^{(1)}(0) + \int_0^t \mathbf{v}^{(1)}(\mathbf{X}^{(1)}(r))\,dr + \mathbf{L}^{(1)}(t) \tag{9}$$

where $\mathbf{v}^{(1)}(\mathbf{X}^{(1)})$ is the Lagrangian velocity, $\mathbf{V}^{(1)}$. In form, (9) is identical to (1), but mathematically $\mathbf{v}^{(1)}(\mathbf{X}^{(1)}(r))$ is rarely stationary. So, the classical central limit arguments do not apply to the asymptotics for the integral in (9). Generally $\mathbf{v}^{(1)}(\mathbf{x})$ has stationary increments in space; and so, we may assume $\mathbf{V}^{(1)}(t)$ will have stationary increments in time. Given a fractal dimension (say, divider dimension) of α_v, we can always construct an α_v-stable Lévy motion with that dimension. Hence, we take $\mathbf{V}^{(1)}(t)$ as α_v-stable Lévy.

There is a subtleness to (9) that separates it from other works (cf. [18,28]). In physics studies concerning the flows in porous media, it is common to assume that the path of the particle is fractal and more specifically Lévy. We do not make this assumption in this chapter. It is more consistent with experimental procedures to assume the Lagrangian velocity, $\mathbf{V}^{(1)}(t)$, rather than the path, $\mathbf{Y}^{(1)}(t)$, is fractal, *i.e.*,

$$\mathbf{Y}^{(1)}(t) = \int_0^t \mathbf{V}^{(1)}(r)dr \tag{10}$$

with $\mathbf{V}^{(1)}$ fractal. Because if we assume the path of the particle is fractal, then its trajectory can be infinitely long during a finite time interval. The exception being matched-index systems [10], one does not observe the path of a particle as it traverses the system. It can be shown [8] that if one bases a definition of fractal on a fractional box dimension, then $\mathbf{Y}^{(1)}(t)$ is not fractal.

Next, consider the renormalization of the convective displacement $\mathbf{Y}^{(1)}$ with $E[\mathbf{V}^{(1)}(t)] = E[\mathbf{V}^{(1)}(0)] = \bar{\mathbf{V}}^{(1)}$ a constant. For fixed t, $\mathbf{V}^{(1)}(t)$ has the α_v-stable distribution: $\mathbf{V}^{(1)}(t) \sim S_{\alpha_v}(tM_v, t\bar{\mathbf{V}}^{(1)})$ [29]. Because of (5), $\mathbf{Y}^{(1)}(t)$ also has an α_v-stable distribution:

$$\mathbf{Y}^{(1)}(t) \sim S_{\alpha_v}\left(\frac{t^{1+\alpha_v}}{1+\alpha_v}M_v, \frac{t^2}{2}\bar{\mathbf{V}}^{(1)}\right).$$

For $t = \lambda t'$ we define $\mathbf{W}_\lambda(t')$ and $\mathbf{Y}^{(2)}(t)$ by

$$\begin{aligned} \mathbf{W}_\lambda(t') &= \frac{\mathbf{Y}^{(1)}(\lambda t') - \lambda t'\bar{\mathbf{V}}^{(1)}}{\lambda^{1+1/\alpha_v}} \\ &= \lambda^{-1/\alpha_v}t'(\mathbf{V}^{(1)}(0) - \bar{\mathbf{V}}^{(1)}) + \lambda^{-(1+1/\alpha_v)}\tilde{\mathbf{Y}}^{(1)}(\lambda t'), \end{aligned} \tag{11}$$

$$\mathbf{Y}^{(2)}(t') = \lim_{\lambda\to\infty} \mathbf{Y}^{(1)}(\lambda t'), \tag{12}$$

respectively. Here, the constant λ provides a renormalized time scale. It is shown [8] that

$$\mathbf{W}_\lambda(t') \xrightarrow{d} \tilde{\mathbf{Y}}^{(2)}(t') \tag{13}$$

which is the renormalization of $\tilde{\mathbf{Y}}^{(1)}$. Convergence in distribution is denoted by $\xrightarrow{d}$. Hence, for any $t > 0$,

$$\mathbf{Y}^{(2)}(t) \approx t\bar{\mathbf{V}}^{(1)} + \tilde{\mathbf{Y}}^{(2)}(t). \tag{14}$$

We now upscale from the mesoscale to the macroscale by taking advantage of the mesoscale periodicity. We have $\mathbf{L}^{(1)}(0) = 0$ and hence $\mathbf{L}^{(1)}(t) \sim S_{\alpha_l}(tM_l, t\mu_l^{(1)})$. Periodicity on the lattice in the velocity is of the form:

$$\mathbf{V}^{(1)}(t; \mathbf{x}_0 + \nu) = \mathbf{V}^{(1)}(t; \mathbf{x}_0), \tag{15}$$

where $\mathbf{x}_0$ is the initial data and $\nu = (\nu_1, \nu_2, \nu_3)$ with ν_j any integer, $j = 1,2,3$. Define the average $[\mathbf{V}^{(1)}]$ by

$$[\mathbf{V}^{(1)}] = E\left[\frac{1}{|A||B|}\int_A\int_B \mathbf{V}^{(1)}(t,\mathbf{x}_0)\,dt\,d\mathbf{x}_0\right] \tag{16}$$

with

$$A = A(\mathbf{x}_0(\omega)) = \{\mathbf{x}_0(\omega) : \mathbf{x}_0(\omega) \in [0,1]^3\} \tag{17}$$

and

$$B = B(t,\omega) = \{t \geq 0 : \mathbf{Y}^{(1)}(t,\omega) \in [0,1]^3\}, \tag{18}$$

where ω is an elementary event.

Define $\mathbf{Z}_\lambda^{(1)}(t)$ by

$$\begin{aligned} \mathbf{Z}_\lambda^{(1)}(t) &= \frac{\tilde{\mathbf{X}}^{(1)}(\lambda t) - \lambda t([\mathbf{V}^{(1)}] + \mu_l^{(1)})}{\lambda^{1+1/\alpha_v}} \\ &= \frac{t(\bar{\mathbf{V}}^{(1)} - [\mathbf{V}^{(1)}] - \mu_l^{(1)})}{\lambda^{1/\alpha_v}} + \frac{\mathbf{Y}^{(1)}(\lambda t) - \lambda t\bar{\mathbf{V}}^{(1)}}{\lambda^{1+1/\alpha_v}} + \frac{\mathbf{L}^{(1)}(\lambda t)}{\lambda^{1+1/\alpha_v}}. \end{aligned} \tag{19}$$

As $\lambda \to \infty$, the first term of the right-hand side converges to zero, the second to $\tilde{\mathbf{Y}}^{(2)}$ and the last to zero in distribution where the reader should recall α_l, $\alpha_v > 1$. Thus, $\mathbf{Z}_\lambda^{(1)}(t) \xrightarrow{d} \tilde{\mathbf{Y}}^{(2)}(t)$. Define $\tilde{\mathbf{X}}^{(2)}(t)$ by $\tilde{\mathbf{X}}^{(2)}(t) = \lim_{\lambda\to\infty} \tilde{\mathbf{X}}^{(1)}(\lambda t)$. Hence, for $t > 0$,

$$\tilde{\mathbf{X}}^{(2)}(t) \approx t([\mathbf{V}^{(1)}] + \mu_l^{(1)}) + \tilde{\mathbf{Y}}^{(2)}(t) \tag{20}$$

which when combined with (14) gives the renormalization for $\tilde{\mathbf{X}}^{(2)}$. Also, since $\tilde{\mathbf{Y}}^{(2)}$ is non-fractal, so is $\tilde{\mathbf{X}}^{(2)}$.

Finally, we discuss the renormalized dispersion (Fokker-Planck) equation for the transition density, f, of $\tilde{\mathbf{X}}^{(2)}$. Let $\hat{f}$ be the Fourier transform of f, then

$$\ln \hat{f}(t,\mathbf{k}) = it\mathbf{v}^{(2)} \cdot \mathbf{k} - \frac{t^{1+\alpha_v}}{1+\alpha_v}\int_{S_3} |\mathbf{k}\cdot\mathbf{s}|^{\alpha_v}\psi(\alpha_v,\mathbf{k},\mathbf{s})M_v(d\mathbf{s}), \tag{21}$$

where $\mathbf{v}^{(2)} = [\mathbf{V}^{(1)}] + \mu_l^{(1)}$ so

$$\begin{aligned} \frac{\partial \hat{f}}{\partial t} &= \mathbf{v}^{(2)} \cdot (i\mathbf{k}\hat{f}) \\ &- \frac{t^{\alpha_v}}{\cos\left(\frac{\pi\alpha_v}{2}\right)}\int_{S_3} |\mathbf{k}\cdot\mathbf{s}|^{\alpha_v}\left(\cos\left(\frac{\pi\alpha_v}{2}\right) - isgn(\mathbf{k}\cdot\mathbf{s})\sin\left(\frac{\pi\alpha_v}{2}\right)\right)M_v(d\mathbf{s})\hat{f}. \end{aligned} \tag{22}$$

The inverse Fourier transform gives

$$\frac{\partial f}{\partial t} = -\mathbf{v}^{(2)} \cdot \nabla f + D^{(2)}(t)\nabla^{\alpha_v} f \tag{23}$$

with $D^{(2)}(t) = -t^{\alpha_v}/\cos(\pi\alpha_v/2) > 0$. Following this theory further [21] we find

$$\begin{aligned} D^{(2)}(t)\nabla^{\alpha_v} f &= \nabla_{\mathbf{x}} \cdot \int_{R^3} (\mathbf{D}^{(2)})'(\mathbf{y},t) \cdot \nabla_{\mathbf{x}-\mathbf{y}} f(\mathbf{x}-\mathbf{y},t)\,d\mathbf{y} \\ &= \nabla_{\mathbf{x}} \cdot \int_0^t \int_{R^3} \tilde{\mathbf{D}}^{(2)}(\mathbf{y},t,\tau) \cdot \nabla_{\mathbf{x}-\mathbf{y}} f(\mathbf{x}-\mathbf{y},t-\tau)\,d\mathbf{y}\,d\tau, \end{aligned} \tag{24}$$

with $1 < \alpha_v \leq 2$ and $\tilde{\mathbf{D}}^{(2)} = (\tilde{D}^{(2)}_{jk})$:

$$\tilde{D}^{(2)}_{jk}(\mathbf{y},t,\tau) = D^{(2)}(t-\tau)\delta(\tau)\mathcal{F}^{-1}\left(\int_{S^3}(i\mathbf{y}\cdot\mathbf{s})^{\alpha_v-2}s_js_kM_v(d\mathbf{s})\right), \tag{25}$$

where $\delta(\tau)$ and $\mathcal{F}^{-1}$ are the Dirac delta and the inverse Fourier transform, respectively.

4. Upscaling from Mesoscale to Macroscale with α-stable Lévy Acceleration

We consider the Lagrangian acceleration to be fractal on the mesoscale rather than the velocity which is the integral of the acceleration. as with the Lagrangian velocity, we assume that the Lagrangian acceleration, $\mathbf{A}^{(1)}$, on the mesoscale is α_a-stable Lévy ($[\mathbf{A}^{(1)}(t) - \mathbf{A}^{(1)}(0)] \sim S_{\alpha_a}(tM_a, t\mu_a)$) with $1 < \alpha_a \leq 2$ and constant mean, $E(\mathbf{A}^{(1)}(t)) = E(\mathbf{A}^{(1)}(0)) = \bar{\mathbf{A}}^{(1)}$ for each $t > 0$. Let

$$\mathbf{Y}^{(1)}(t) = \int_0^t \int_0^u \mathbf{A}^{(1)}(r)drdu.$$

We decompose $\mathbf{Y}^{(1)}(t)$ as follows:

$$\mathbf{Y}^{(1)}(t) = \int_0^t \int_0^u \left(\mathbf{A}^{(1)}(0) + \tilde{\mathbf{A}}^{(1)}(r)\right) dr\,du = \frac{t^2}{2}\mathbf{A}^{(1)}(0) + \tilde{\mathbf{Y}}^{(1)}(t),$$

where

$$\tilde{\mathbf{Y}}^{(1)}(t) = \int_0^t \int_0^u \tilde{\mathbf{A}}^{(1)}(r)drdu \tag{26}$$

and $\mu_a = (1/r)E(\tilde{\mathbf{A}}^{(1)}(r)) = \mathbf{0}$. By switching the two integrals in (26), we obtain

$$\tilde{\mathbf{Y}}^{(1)}(t) = \int_0^t (t-r)\tilde{\mathbf{A}}^{(1)}(r)dr.$$

For each t, the characteristic function, ϕ_y, of $\tilde{\mathbf{Y}}(t)$ is

$$\phi_y(t,\mathbf{k}) = \exp\left[-\frac{t^{1+2\alpha_a}}{2^{\alpha_a}(1+2\alpha_a)}\int_{S_3}|\mathbf{k}\cdot\mathbf{s}|^{\alpha_a}\psi(\alpha_a,\mathbf{k},\mathbf{s})M_a(d\mathbf{s})\right]. \tag{27}$$

We can prove (27) [14] in a fashion similar to (5).

We take the same assumption on the microscale as in the previous section. and upscale from mesoscale to macroscale. We assume periodicity of $\mathbf{A}^{(1)}(r) \equiv \mathbf{A}^{(1)}(r,\xi^{(1)}(0))$ in the initial data:

$$\mathbf{A}^{(1)}(r,\xi^{(1)}(0) + \nu^{(1)}) = \mathbf{A}^{(1)}(r,\xi^{(1)}(0)),$$

where $\nu^{(1)} = (\nu^{(1)}_1, \cdots, \nu^{(1)}_d)$ for integers $\nu^{(1)}_k$, $k = 1, \cdots, d$ and $\xi^{(1)}(\cdot) = \xi^{(1)}(\cdot,\omega)$ is a particle path in the porous medium. Let

$$[\mathbf{A}^{(1)}] = E\left[\frac{1}{|\Omega_0|}\int_{\Omega_0}\frac{1}{|B|}\int_B \mathbf{A}^{(1)}(t,\xi^{(1)}(0))\,dt\,d\xi^{(1)}(0)\right], \tag{28}$$

where $\Omega_0 = \Omega_0(\omega) = \{\xi^{(1)}(0,\omega) : \xi^{(1)}(0,\omega) \in [0,1]^d\}$ and $B = B(\omega) = \{t \geq 0 : \xi^{(1)}(t,\omega) \in [0,1]^d\}$. Set $\mathbf{L}^{(1)}(t) = t\mu_\ell^{(1)} + \tilde{\mathbf{L}}^{(1)}(t)$.

For $t = \lambda t'$, we can show the convergence by the same argument as in (11):

$$\frac{\mathbf{Y}^{(1)}(\lambda t') - \frac{1}{2}(\lambda t')^2\bar{\mathbf{A}}^{(1)}}{\lambda^{2+1/\alpha_a}} \xrightarrow{d} \mathcal{W}(t'), \qquad \lambda \to \infty.$$

The characteristic function, ϕ_w, of $\mathcal{W}(t')$ is the same as (27). Let $\tilde{\mathbf{Y}}^{(2)}(t') = \lim_{\lambda\to\infty} \tilde{\mathbf{Y}}^{(1)}(\lambda t')$. Thus, for each $t > 0$, we have

$$\mathbf{Y}^{(2)}(t) \approx \frac{t^2}{2}\bar{\mathbf{A}}^{(1)} + \tilde{\mathbf{Y}}^{(2)}(t). \tag{29}$$

To obtain a central limit theorem for $\tilde{\mathbf{X}}^{(1)}(t)$, let

$$\begin{aligned} \mathcal{Z}_\lambda^{(1)}(t') &\equiv \frac{\tilde{\mathbf{X}}^{(1)}(\lambda t') - \frac{(\lambda t')^2}{2}[\mathbf{A}^{(1)}] - \lambda t'\mu_\ell^{(1)}}{\lambda^{2+1/\alpha_a}} \\ &= \frac{(t')^2(\bar{\mathbf{A}}^{(1)} - [\mathbf{A}^{(1)}])}{2\lambda^{1/\alpha_a}} + \frac{\mathbf{Y}^{(1)}(\lambda t') - \frac{1}{2}(\lambda t')^2\bar{\mathbf{A}}^{(1)}}{\lambda^{2+1/\alpha_a}} \\ &+ \frac{\rho^{(1)}\mathbf{L}^{(1)}(\lambda t') - \lambda t'\mu_\ell^{(1)}}{\lambda^{2+1/\alpha_a}}. \end{aligned} \tag{30}$$

As $\lambda \to \infty$, the first and second terms of the right-hand side of (30) converge to zero and $\tilde{\mathbf{Y}}^{(2)}(t')$ in distribution, respectively. By the monotone convergence theorem, the last term of the right-hand side of (30) converges to zero in distribution as $\lambda \to \infty$. So, $\mathcal{Z}_\lambda^{(1)}(t')$ converges to $\tilde{\mathbf{Y}}^{(2)}(t')$ in distribution. Thus, for each $t > 0$, we can approximate the stochastic process $\tilde{\mathbf{X}}^{(2)}(t) = \lim_{\lambda\to\infty} \tilde{\mathbf{X}}^{(1)}(\lambda t)$ as

$$\tilde{\mathbf{X}}^{(2)}(t) \approx \left(\frac{t^2}{2}[\mathbf{A}^{(1)}] + t\mu_\ell^{(1)}\right) + \tilde{\mathbf{Y}}^{(2)}(t). \tag{31}$$

It is possible to derive the Fokker-Planck (Advection-Dispersion) equation [8] that the transition density $f(t,\mathbf{x})$ for $\tilde{\mathbf{X}}^{(2)}(t)$ satisfies:

$$\frac{\partial f}{\partial t} = -\mathbf{v}^{(2)} \cdot \nabla f + D^{(2)}(t)\nabla^{\alpha_a} f, \tag{32}$$

where $\mathbf{v}^{(2)} = t[\mathbf{A}^{(1)}]/2 + \mu_\ell^{(1)}$ and $D^{(2)}(t) = -t^{2\alpha_a}/\cos(\pi\alpha_a/2)$.

5. Finite-size Lyapunov Exponent

In order to apply an α-stable Lévy motion to bacterial motility, one has to choose the stability index α as exactly as possible. As a simple method for determining the index α, we introduce the finite-size Lyapunov exponent (FSLE) [30-33] which is the exponential rate at which two particles separate from a distance of r to $a \times r$ ($a > 1$) and provides a measure of dispersive mixing in chaotic systems. We follow this study as shown in [33]

The concept of a FSLE has been applied to study dispersion in turbulent flow fields [30-32,34-36], dispersion in porous systems [32] the relative dispersion on the surface of the Gulf of Mexico [37], the mixing structures in the Mediterranean Sea [38] and the transport of balloons in the lower stratosphere [39].

A rigorous statistical mechanical definition of the FSLE was presented in [32] that takes into account such phenomena as recirculation, nonstationarity, and periodic processes. If N is the total number of particles in a completely expansive system and the joint probability measure that particles i and j have separation in $(r_{i,j}, r_{i,j}+dr_{i,j})$ at time τ and separation in $(r'_{i,j}, r'_{i,j}+dr'_{i,j})$ at time $t+\tau$ is denoted by $f(r_{1,2}(\tau), r_{1,3}(\tau), \cdots, r_{N-1,N}(\tau), r'_{1,2}(t+\tau), r'_{1,3}(t+\tau), \cdots, r'_{N-1,N}(t+\tau))dr_{1,2}dr'_{1,2}\cdots dr_{N-1,N}dr'_{N-1,N}$ in a $[N^2(N-1)^2]$-dimensional pair separation space $\Omega(t,\tau) = \Omega(\tau)\times\Omega(t+\tau)$, then the conditional probability of any two particles being separated by a distance $a\times x$ for the first time at $t+\tau$, given they were separated by x at time τ, is expressed as

$$G_\tau(x, ax; t) = \frac{1}{N(N-1)}\sum_{i\neq j}\frac{\ll \delta(x-r_{i,j}(\tau))\delta(ax-r_{i,j}(t+\tau))\gg}{\ll \delta(x-r_{i,j}(\tau))\gg}, \tag{33}$$

where δ is the usual Dirac delta and $\ll\cdot\gg$ indicates integration over $\Omega(t,\tau) = \Omega(\tau)\times\Omega(t+\tau)$ with respect to the joint probability density as defined earlier. The average conditional probability over all pairs of particles to reach separation of $a\times r$ for the first time at t, given that they were separated by r at time zero, can be found by integrating (33) for all possible value of τ as

$$G(r, ar; t) = \int_0^\infty G_\tau(r, ar; t)d\tau. \tag{34}$$

If r is the initial separation between two trajectories and $T_a(r)$ is the mean time it takes the separation to grow by a factor $a > 1$, then one can write

$$T_a(r) = E[G(r, ar; t)] = \int_0^\infty tG(r, ar; t)dt. \tag{35}$$

The FSLE of a diverging system is defined as

$$\lambda_a(r) = \frac{1}{T_a(r)}\ln a. \tag{36}$$

This has been shown computationally and theoretically in [32] and [33] respectively, that

$$\lambda_a(r) \sim r^{-\alpha}. \tag{37}$$

Therefore, the stability parameter α of an α-stable Lévy process can be determined by measuring the time $T_a(r)$ using any available particle-tracking algorithm [10]

6. Renormalizing Operator-stable Lévy Motions in Anisotropic Porous Media

While microbes can propel themselves by many means, the driving mechanism most commonly studied [9] is a series of rotary motors that spin flagella phase in a common direction. Periodically, one of these motors will reverse direction and its flagella will entangle the remaining flagella thus stopping and rotating the bug. After the motors have realigned themselves, the microbe heads off in a random direction. Runs are generally longer in the direction of an energy (light, chemical, heat, magnetic) source, and thus affectively the microbe appears to swim towards its food. The resultant swimming process looks like a random walk in 3-dimensions. As shown in [14,15], we assume the motility of a microbe can be described by an operator-stable Lévy motion on the pore (micro) scale. This assumption allows the bug to preferentially run down energy gradients within the pore. Lévy processes are characterized by very heavy tails with infinite second moment. In addition, their probability densities generally have to be defined via their characteristic functions (i.e., the Fourier transform of the probability density) and consequently most of the subsequent analysis is carried out in Fourier space. The mathematical theory of operator-stable Lévy motions can be found in [11-13,40] The operator-stable Lévy diffusion is coupled with a microscale drift that represents the mean pore water velocity, which is assumed stationary, ergodic, and Markovian.

We restrict the analysis to the non-absorptive case, but generalize the approaches in previous sections to handle anisotropic porous media. This generalization is straight forward, but is of critical importance, because almost all natural porous media are anisotropic in 3-dimensions. The main difference between the approach presented here and the previous analysis is that here we allow components of the mesoscale Lagrangian velocity to have different fractal dimensions in the principle directions of flow. This affectively allows for the representation of fractal anisotropy in the Darcy conductivity. Specifically, we assume dispersion on the mesoscale (the regional homogenization of pore and solid space) is represented by the asymptotics of the microscale convection plus pore diffusion, coupled to the Lagrangian Darcy drift velocity.

An infinitely divisible distribution μ [17] on $\mathbf{R}^d$ is called operator-stable [11,12,17] if it has the property that for any $a > 0$, there is an invertible $d \times d$ matrix $Q = a^G$ and $\mathbf{c} \in \mathbf{R}^d$ such that

$$\hat{\mu}(\mathbf{k})^a = \hat{\mu}(Q^T\mathbf{k})e^{i\mathbf{c}\cdot\mathbf{k}}, \tag{38}$$

where a^G is defined by the series $\sum_{m=0}^{\infty}(\ln a)^m G^m/(m!)$, Q^T is the transpose of the matrix Q, and $\hat{\mu}(\mathbf{k})$ is the characteristic function of μ:

$$\hat{\mu}(\mathbf{k}) = \int_{\mathbf{R}^d} e^{i\mathbf{k}\cdot\mathbf{x}}\mu(d\mathbf{x}), \qquad \mathbf{k} \in \mathbf{R}^d.$$

The matrix G is called an exponent for μ. If $G = (1/\alpha)I$ with the $d \times d$ identity matrix I, then μ is a d-dimensional α-stable distribution. A stochastic process $\{\mathbf{V}(t)|t \geq 0\}$ is called an operator-stable Lévy process if it is a Lévy process and its increments have operator-stable distributions. Because of (38), an operator-stable Lévy process $\mathbf{V}(t)$ has a scaling law: for any $a > 0$ and $t > 0$, there is a $d \times d$ matrix G and $\mathbf{c}(t) \in \mathbf{R}^d$, such that

$$\mathbf{V}(at) \stackrel{d}{=} a^G\mathbf{V}(t) + \mathbf{c}(t), \tag{39}$$

where $\stackrel{d}{=}$ means equal in distribution.

Consider the following integrated SODE in three dimensions:

$$\mathbf{X}(t) = \mathbf{X}(0) + \int_0^t \mathbf{V}(r)\,dr + \mathbf{L}(t), \quad t \geq 0, \tag{40}$$

where $\{\mathbf{V}(r)\}$ and $\{\mathbf{L}(t)\}$ are operator-stable Lévy processes. Subsequently, this will be our model of microbial dynamics at the mesoscale for an anisotropic fractal porous medium. We present a brief analysis of this Lagrangian equation and take a look at its Eulerian counterpart. The characteristic function ϕ_v [12] of $\tilde{\mathbf{V}}(t) \sim (t\mathcal{A}_v, t\nu_v, t\mu_v)$, called the generating triplet [17] of ϕ_v, is

$$\phi_v(t,\mathbf{k}) = \exp\left[-\frac{t}{2}\mathbf{k}\cdot\mathcal{A}_v\cdot\mathbf{k} + t\hat{B}(\mathbf{k}) + it\mathbf{k}\cdot\mu_v\right], \tag{41}$$

where $t\mu_v = E[\tilde{\mathbf{V}}(t)]$ is the expected value of $\tilde{\mathbf{V}}(t)$, $\mathcal{A}_v$ a symmetric nonnegative definite 3×3 matrix, and

$$\hat{B}(\mathbf{k}) = \int_{S_3}\int_0^\infty \left(e^{i\mathbf{k}\cdot(r^G\mathbf{s})} - 1 - i\mathbf{k}\cdot(r^G\mathbf{s})\right)\frac{dr}{r^2}M_v(d\mathbf{s}). \tag{42}$$

We will use the index, ℓ, to represent the characteristic function of $(\mathbf{L}(t)-\mathbf{L}(0))$. If $M_v = 0$, then $\tilde{\mathbf{V}}(t)$ is Brownian, which is a special case in [8]. We assume $M_v \neq 0$ in this section. Following these results [8], we compute the characteristic function $\phi_y(t,\mathbf{k}) = E[\exp(i\mathbf{k}\cdot\tilde{\mathbf{Y}}(t))]$ of $\tilde{\mathbf{Y}}(t)$:

$$\phi_y(t,\mathbf{k}) = \exp\left[-\frac{t^3}{6}\mathbf{k}\cdot\mathcal{A}_v\cdot\mathbf{k} + t\hat{H}(t,\mathbf{k}) + \frac{t^2}{2}i\mathbf{k}\cdot\mu_v\right], \tag{43}$$

where

$$\hat{H}(t,\mathbf{k}) = \int_{S_3}\int_0^\infty \left(\frac{e^{it\mathbf{k}\cdot(r^G\mathbf{s})} - 1}{it\mathbf{k}\cdot(r^G\mathbf{s})} - 1 - \frac{1}{2}it\mathbf{k}\cdot(r^G\mathbf{s})\right)\frac{dr}{r^2}M_v(d\mathbf{s}). \tag{44}$$

6.1. Upscaling from Microscale to Mesoscale

In this microscale, we assume the behavior of microbes depends on direction and so the integrated SODE is of the form:

$$\mathbf{X}^{(0)}(t) = \mathbf{X}^{(0)}(0) + \int_0^t \mathbf{V}^{(0)}(r)\,dr + \mathbf{L}^{(0)}(t), \quad t \geq 0, \tag{45}$$

where $\mathbf{V}^{(0)}(r)$ is assumed stationary, ergodic and Markovian with the mean $\bar{\mathbf{V}}^{(0)}$, and $\mathbf{L}^{(0)}(t)$ is an operator-stable Lévy process $(\mathbf{L}^{(0)}(t) \sim (t\mathcal{A}_\ell^{(0)}, t\nu_\ell^{(0)}, t\mu_\ell^{(0)}))$ which corresponds to the microbes' motility.

The following convergence has been proved in [14]:

$$\frac{\tilde{\mathbf{X}}^{(0)}(\lambda t) - \lambda t(\bar{\mathbf{V}}^{(0)} + \mu_\ell^{(0)})}{\lambda^{1/2}} \xrightarrow{d} \tilde{\mathbf{L}}^{(1)}(t) \tag{46}$$

whose characteristic function, ψ_ℓ, is

$$\psi_\ell(t,\mathbf{k}) = \exp\left[-\frac{t}{2}\mathbf{k}\cdot\mathcal{A}_\ell^{(1)}\cdot\mathbf{k} - \frac{t}{2}\int_{S_3}\int_0^\infty (\mathbf{k}\cdot(r^G\mathbf{s}))^2\frac{dr}{r^2}M_\ell(d\mathbf{s})\right]. \tag{47}$$

6.2. Upscaling Operator-stable Lévy Drift Velocity

Here, we consider the case where the porous medium induces a fractal velocity field. Non-fractal velocity is upscaled as in [15]. Let $\mathbf{V}^{(1)}(t) = (V_1^{(1)}(t), V_2^{(1)}(t), V_3^{(1)}(t))^T$ be an operator-stable Lévy process with constant mean, $E(\mathbf{V}^{(1)}(t)) = \bar{\mathbf{V}}^{(1)}$ for any $t \geq 0$ and an exponent $G = \mathrm{diag}(1/\alpha_1, 1/\alpha_2, 1/\alpha_3)$. This process was studied in [41-43] Hendricks [41,42] and Pruitt and Taylor [43] considered the case where the components $V_1^{(1)}(t), V_2^{(1)}(t), V_3^{(1)}(t)$ are independent.

We assume $1 < \alpha_j \leq 2$ for $j = 1,2,3$, $\alpha_o \equiv \min\{\alpha_j | j = 1,2,3\} < 2$, and the components, $V_1^{(1)}(t), V_2^{(1)}(t), V_3^{(1)}(t)$, are independent. Each component $V_j^{(1)}(t)$ is an α_j-stable Lévy process with divider dimension α_j. The characteristic function of $\tilde{\mathbf{V}}^{(1)}(t) = (\tilde{V}_1^{(1)}(t), \tilde{V}_2^{(1)}(t), \tilde{V}_3^{(1)}(t))^T$ is

$$\begin{aligned} E\left[\exp(i\mathbf{k}\cdot\tilde{\mathbf{V}}^{(1)}(t))\right] &= E\left[\exp(i\sum_{j=1}^{3} k_j \tilde{V}^{(1)}(t))\right] = \prod_{j=1}^{3} E\left[\exp(ik_j\tilde{V}^{(1)}(t))\right] \\ &= \prod_{j=1}^{3} \exp\left[-t\sigma_j^{\alpha_j}|k_j|^{\alpha_j}\left(1 - i\beta_j sgn(k_j)\tan\left(\frac{\pi\alpha_j}{2}\right)\right)\right]. \end{aligned}$$

Note, that $\tilde{\mathbf{V}}^{(1)}(t)$ has infinite second moment [17] because $\alpha_o < 2$. We can compute the natural logarithm of the characteristic function of $\tilde{\mathbf{Y}}^{(1)}(t)$ in the same fashion as in section 2:

$$\ln E[\exp(i\mathbf{k}\cdot\tilde{\mathbf{Y}}^{(1)}(t))] = \sum_{j=1}^{3}\left[-\frac{t^{1+\alpha_j}\sigma_j^{\alpha_j}}{1+\alpha_j}|k_j|^{\alpha_j}\left(1 - i\beta_j sgn(k_j)\tan\left(\frac{\pi\alpha_j}{2}\right)\right)\right].$$

Letting $t = \lambda t'$, then we can show that

$$\begin{aligned} \frac{\mathbf{Y}^{(1)}(\lambda t') - \lambda t'\bar{\mathbf{V}}^{(1)}}{\lambda^{1+1/\alpha_o}} &= \lambda^{-1/\alpha_o}t'(\mathbf{V}^{(1)}(0) - \bar{\mathbf{V}}^{(1)}) + \lambda^{-1-1/\alpha_o}\tilde{\mathbf{Y}}^{(1)}(\lambda t') \\ &\xrightarrow{d} \Xi(t'), \end{aligned} \tag{48}$$

as $\lambda \to \infty$, where the natural log of the characteristic function ϕ_Ξ of $\Xi(t')$ is

$$\ln\phi_\Xi(t',\mathbf{k}) = \sum_{j\in I_o}\left[-\frac{t^{1+\alpha_j}\sigma_j^{\alpha_j}}{1+\alpha_j}|k_j|^{\alpha_j}\left(1 - i\beta_j sgn(k_j)\tan\left(\frac{\pi\alpha_j}{2}\right)\right)\right] \tag{49}$$

with $I_o = \{j \,|\, \alpha_j = \alpha_o,\ j = 1,2,3\}$. Let $\mathbf{Y}^{(2)}(t) = \lim_{\lambda\to\infty}\mathbf{Y}^{(1)}(\lambda t)$, then for $t > 0$ we have

$$\mathbf{Y}^{(2)}(t) \approx t\bar{\mathbf{V}}^{(1)} + \Xi(t). \tag{50}$$

We also assume periodicity of $\mathbf{V}^{(1)}(t) \equiv \mathbf{V}^{(1)}(t, \xi^{(1)}(0))$ in the initial data and use

$$[\mathbf{V}^{(1)}] = E\left[\frac{1}{|\Omega_0|}\int_{\Omega_0}\frac{1}{|A|}\int_A \mathbf{V}^{(1)}(t,\xi^{(1)}(0))\,dt\,d\xi^{(1)}(0)\right],$$

to upscale from mesoscale to macroscale. Let

$$\begin{aligned}\mathbf{Z}_{\lambda}^{(1)}(t') &\equiv \frac{\tilde{\mathbf{X}}^{(1)}(\lambda t') - ([\mathbf{V}^{(1)}] + \mu_{\ell}^{(1)})\lambda t'}{\lambda^{1+1/\alpha_o}} \\ &= \frac{t'(\bar{\mathbf{V}}^{(1)} - [\mathbf{V}^{(1)}])}{\lambda^{1/\alpha_o}} + \frac{\mathbf{Y}^{(1)}(\lambda t') - \lambda t' \bar{\mathbf{V}}^{(1)}}{\lambda^{1+1/\alpha_o}} \\ &+ \frac{\mathbf{L}^{(1)}(\lambda t') - \lambda t' \mu_{\ell}^{(1)}}{\lambda^{1+1/\alpha_o}}.\end{aligned}$$

Therefore, $\mathbf{Z}_{\lambda}^{(1)}(t')$ converges to $\Xi(t')$ in distribution. For each $t > 0$, we can approximate the stochastic process which we label $\tilde{\mathbf{X}}^{(2)}(t) = \lim_{\lambda\to\infty} \tilde{\mathbf{X}}^{(1)}(\lambda t)$ as

$$\tilde{\mathbf{X}}^{(2)}(t) \approx t([\mathbf{V}^{(1)}] + \mu_{\ell}^{(1)}) + \Xi(t). \tag{51}$$

If $\beta_j = 1$ for $j \in I_o$, then we can derive from (49) the fractional Fokker-Planck equation that the transition density $f(t, \mathbf{x})$ for $\tilde{\mathbf{X}}^{(2)}(t)$ satisfies

$$\frac{\partial f}{\partial t} = -\mathbf{v}^{(2)} \cdot \nabla f + \sum_{j\in I_o} D_j^{(2)}(t) \frac{\partial^{\alpha_j} f}{\partial x_j^{\alpha_j}}, \tag{52}$$

where $\mathbf{v}^{(2)} = [\mathbf{V}^{(1)}] + \mu_{\ell}^{(1)}$, $D_j^{(2)}(t) = -t^{\alpha_j} \sigma_j^{\alpha_j} / \cos(\pi\alpha_j/2)$, and $\frac{\partial^{\alpha} f}{\partial x_j^{\alpha}}$ is defined by

$$\widehat{\frac{\partial^{\alpha} f}{\partial x_j^{\alpha}}}(t, \mathbf{k}) = \left[|-k_j|^{\alpha} \exp\left(i sgn(-k_j)\frac{\pi\alpha}{2}\right)\right] \hat{f}(t, \mathbf{k}). \tag{53}$$

The fractional order is the divider dimension of the velocity process in the principle directions.

Remark 6.1. *Our approach can be extended to an operator-stable Lévy process* $\mathbf{V}(t)$ *with a diagonalizable exponent* $G = P\mathcal{D}P^{-1}$, *where* P *is an invertible matrix and* $\mathcal{D} = diag(1/\alpha_1, 1/\alpha_2, 1/\alpha_3)$, *if* $P^{-1}\mathbf{V}(t)$ *has independent components. Because for any* $a > 0$ *and* $t > 0$,

$$P^{-1}\mathbf{V}(at) \stackrel{d}{=} a^{\mathcal{D}} P^{-1}\mathbf{V}(t) + P^{-1}\mathbf{c}(t). \tag{54}$$

Remark 6.2. *When* $\mathbf{L}^{(0)}(t) = (L_1^{(0)}(t), L_2^{(0)}(t), L_3^{(0)}(t))^T$ *is an operator-stable Lévy process with an exponent* $G_l = diag(1/\alpha_1^l, 1/\alpha_2^l, 1/\alpha_3^l)$ $(1 < \alpha_j^l \le 2,\ j = 1,2,3)$ *and independent components at the microscale we can also upscale* $\mathbf{X}^{(0)}(t)$ *in (45) with a stationary, ergodic, and Markovian process,* $\mathbf{V}^{(0)}(t)$, *with the mean* $\bar{\mathbf{V}}^{(0)}$ *from microscale to mesoscale and from mesoscale to macroscale in the same fashion, and obtain the same Fokker-Planck equation (52).*

Conclusion

To summarize, we have been studying the evolution of genetic information in microbial communities in natural porous systems. One facet of this complicated problem is modeling the evolution of microbes as they traverse a porous body. More specifically, we are interested in the "large" scale statistics of the process. To rationally address this problem, a method to renormalize the trajectories is required. The purpose of this chapter was to show how this can be accomplished via central limit theorems for integrated stochastic ODE's with Lévy drift and Lévy diffusion. The Lévy diffusion represents the motility of the individual microbe on the pore scale while the Lévy drift characterizes the fractal nature of the convective field on the mesoscale. The renormalization procedure is of sufficient generality to be applicable to the transport of any colloid particle in a porous body that displays a fractal velocity field on some scale. The renormalized dispersion equations are fractional with time-dependent positive-definite dispersion coefficients. An explicit representation of the dispersion coefficients was provided.

To model all possible microbes' swimming processes and anisotropic porous media the approach was generalized by adopting operator-stable Lévy motions. The operator-stable Levy diffusion was coupled with a microscale drift that represents the pore water velocity, which was assumed stationary, ergodic, and Markovian. The main difference of the approach in section 6 from that in section 4 is that we allow components of the mesoscale Lagrangian velocity to have different fractal dimensions in the principle directions of flow. This affectively allows for the representation of fractal anisotropy in the Darcy conductivity. Specifically, we assumed dispersion on the mesoscale is represented by the asymptotics of the microscale convection plus pore diffusion, coupled to the Lagrangian Darcy drift velocity.

The macroscale porous formation was assumed periodic and transport of the colloid on this scale was given by the renormalization of the mesoscale drift with the mesoscale dispersive flux. There was no macroscale drift that is distinct from the macroscale dispersive flux.

Fokker-Planck equations were obtained on each of the three scales considered. This was accomplished by first deriving the SODE on the meso- and macro-scales via application of generalized central limit theorems. This approach can be shown equivalent to a rigorous renormalization group approach. The Fokker-Planck equations are in general fractional with different fractal dimensions in the principle directions.

References

[1] H. Nikaido, *Ann. Rev. Biochem.*, 78, 119 (2009).

[2] NIAID Tuberculosis Workiing Group, report, 2007 at URL: http://www.niaid.nih.gov/topics/tuberculosis/Research/Documents/mdrxdrresearchagenda.pdf.

[3] S. Mathur and R. Singh, *Int. J. Food Micro.*, 105, 281 (2005).

[4] J.W. Costerton, Z. Lewandowski, D.E. Caldwell, D.R. Korber, and H.M. Lappin-Scott, *Ann. Rev. Microbiol.*, 49, 711 (1995).

[5] R.M. Harshey, *Ann. Rev. Microbiol.*, 57, 249 (2003).

[6] B.B. Mandelbrot, *The Fractal Geometry of Nature*, W.H. Freeman, New York (1983).

[7] K. Falconer, *Fractal Geometry: Mathematical Foundations and Applications*, Second Edition, Wiley, New York (2003).

[8] M. Park, N. Kleinfelter, and J.H. Cushman, *SIAM Multiscale Mod. Sim.*, 4, 4, 1233 (2005).

[9] H.C. Berg, *Phys. Today*, 53, 1, 24 (2000).

[10] M. Moroni and J.H. Cushman, *Phys. Fluids*, 13, 1, 81 (2001).

[11] M. Sharpe, *Trans. Amer. Math. Soc.*, 136, 51 (1969).

[12] Z.J. Jurek and J.D. Mason, *Operator-Limit Distributions in Probability Theory*, Wiley, New York (1993).

[13] M.M. Meerschaert and H.P. Scheffler, *Handbook of Heavy Tailed Distributions in Finance*, Edited by S.T. Rachev, Elsevier Science, Amsterdam, 595 (2003).

[14] M. Park and J.H. Cushman, *J. Comp. Phys.*, 217, 159 (2006).

[15] M. Park and J.H. Cushman, *J. Stat. Mech.*, 2009, 2, P02039 (2009).

[16] A. Janicki and A. Weron, *Simulation and Chaotic behavior of α-stable Stochastic processes*, Marcel Dekker, New York (1994).

[17] K. Sato, *Lévy processes and infinitely divisible distributions*, Cambridge University Press, Cambridge (1999).

[18] M.M. Meerschaert, D.A. Benson, and B. Baeumer, *Phys. Rev. E*, 59, 5026 (1999).

[19] I. Karatzas and S.E. Shreve, *Brownian Motion and Stochastic Calculus*, Springer, Second edition (1991).

[20] L. Ahlfors, *Complex Analysis*, McGraw-Hill Science/Engineering/Math; 3 edition (1979).

[21] J.H. Cushman and M. Moroni, *Phys. Fluids*, 13, 1, 75 (2001).

[22] J.H. Cushman, M. Park, N. Kleinfelter, and M. Moroni, *Geophy. Res. Lett.*, 32, L19816 (2005).

[23] M. Cáceres, *Braz. J. Phys.*, 29, 1, 125 (1999).

[24] M. Park, N. Kleinfelter, and J.H. Cushman, *Geoph. Res. Lett.*, 33, L01401 (2006).

[25] R.N. Bhattacharya, *Z. Wahrsch. verw. Gebiete*, 60, 185 (1982).

[26] R.N. Bhattacharya, *Ann. Prob.*, 13, 385 (1985).

[27] R.N. Bhattacharya and V.K. Gupta, in *Dynamics of Fluids in Hierarchical Porous Media*, edited by Cushman, J.H., Academic Press, London, 61 (1990).

[28] M.M. Meerschaert, D.A. Benson, and B. Baeumer, *Phys. Rev. E*, 63, 021112 (2001).

[29] G. Samorodnitsky and M.S. Taqqu, *Stable Non-Gaussian Random Processes: Stochastic Models with Infinite Variance*, Chapman and Hall, New York (1994).

[30] E. Aurell, G. Boffetta, A. Crisanti, G. Paladin, and A. Vulpiani, *Phys. Rev. Lett.*, 77, 1262 (1996).

[31] E. Aurell, G. Boffetta, A. Crisanti, G. Paladin, and A. Vulpiani, *Physica A*, 30, 1 (1997).

[32] N. Kleinfelter, M. Moroni, and J.H. Cushman, *Phys. Rev. E.*, 72, 056306 (2005).

[33] R. Parashar and J.H. Cushman, *Phys. Rev. E*, 76, 017201 (2007).

[34] V. Artale, G. Boffetta, A. Celani, M. Cencini, and A. Vulpiani, *Phys. Fluids*, 9, 3162 (1997).

[35] G. Boffetta, M. Celani, M. Cencini, G. Lacorata, and A. Vulpiani, *Chaos*, 10, 50 (2000).

[36] G. Boffetta, M. Cencini, S. Espa, and G. Querzoli, *Europhys. Lett.*, 48, 629 (1999).

[37] J.H. La Casce and C. Ohlmann, *J. Mar. Res.*, 61, 285 (2003).

[38] F. d'Ovidio, V. Fernandez, E. Hernandez-Garcia, and C. Lopez, *Geophys. Res. Lett.*, 31, L17203 (2004).

[39] G. Lacorata, E. Aurell, B. Legras, and A. Vulpiani, *J. Atmos. Sci.*, 61, 2936 (2004).

[40] S. Resnick and P. Greenwood, *J. Multivariate Anal.*, 9, 206 (1979).

[41] W.J. Hendricks, *Z. Wahrsch. Verw. Gebiete*, 28, 113 (1974).

[42] W.J. Hendricks, *Z. Wahrsch. Verw. Gebiete*, 49, 13 (1979).

[43] W.E. Pruitt and S.J. Tayor, *Z. Wahrsch. Verw. Gebiete*, 12, 267 (1969).

In: Medical Physics
Editor: Marcin Balcerzyk

ISBN: 978-1-62257-590-9

Chapter III

Color Vision and Wavefront Aberrations of the Human Eye

Florin Toadere*
National Institute for Research and Development of Isotopic and Molecular Technologies,Cluj-Napoca, Romania

Abstract

The visual system of the human eye is a part of the central nervous system, by which the human body sees and interprets the information provided by the visible light in order to build a representation of the world around. The nervous system derives the colors by comparing the responses to light of the three types of photodetectors cones. In the case of normal vision, the cones are divided in three categories: long, medium and short, which are sensitive to the blue, green and red zones of the visible spectrum. For people who are color blind, the task of distinguishing different colors is very difficult, because in their eyes exist cones' disorder, or a category of cones is missing. We simulate the daylight perception by a normal eye and an eye of the Daltonist. In order to achieve a good visual performance of the eye, it needs to have good visual acuity and a good sensitivity to the contrast.

In addition to color perception, human eye can suffer from diseases and disorders. An eye can be seen as an optical instrument, which is characterized by its unique eye print. The human eye is an imperfect optical instrument, and there are not two individuals with eye print exactly alike, even for their two eyes. The ocular aberrations can be considered as a combination of the spherocylindrical aberrations and irregular aberrations, which are called low-order aberrations and high-order aberration. We analyze some aspects related to methods and technologies thatcan be used to describe the wavefronts aberrations of the human eye. We use the formula of the Siedel aberrations and the Zenike polynomials in order to determine the wavefront aberrations of the human eye. We analyze the principle of functionality of some nowadays apparatus and technologies that can be used to capture the wavefront aberrations of the eye. We analyze the Shack-Hartmann sensor, the interferometer aberrations fringes and the Talbot imaging. Finally, to remove the visual aberrations and to improve the acuity of the eye,

* E-mail address: florin.toadere@itim-cj.ro.

we briefly present the correction of the effect of the aberrations using the LASIK (laser assisted in situ keratomileusis) surgery.

1. Introduction

Human vision is sensitive to visible light, that part of the electromagnetic spectrum thatcontains wavelengths between 400 nm and 700 nm. Light can be understood as the aspect of radiant energy thatan observer perceives through visual sensation but also as a physical phenomenon [7,12,15,22,33]. The visual system is the detector that transforms radiant power into luminous sensation. The imaging process accounts for the spatial light distribution on the retina by the optical eye media (cornea, pupil, lens, vitreous, etc.). In this chapter, we describe some principles of human vision thatcan be used to simulate the functionality of the eye. We present aspects related to color vision and some methods to determine and correct the human eye aberrations.

The illumination determines the amount of light that covers a surface. Color helps in the perception of the beauty of a landscape or of a digital image. The perceived color of the surface is determined not only by the color of the surface but also by the color of the light. Therefore, the type of lamp in the light fixture can significantly impact the perception of the color of the object. This effect should be taken into account when we see images on the computer screen. The perception of the object's shape differs with the light distribution on its surface and with the configuration of the resulting shadows. The direction of the light beam can easily affect the perception of the object's shape [12,17,21-23,26,29-31].

Therefore, the daylight’s color temperature during the phase of the day can significantly impact the perception of the color of the object. Radiation from the sun that does get through the atmosphere and is visible to our eyes can be described in three ways: sunlight, skylight and daylight [7,12,15].

- Sunlight is light coming directly from the sun; the image of a shaft of sunlight into a darkened room.
- Skylight refers to the blue light of the sky as viewed from a location in complete shade, for example, the light entering through a north-facing window.
- Daylight is the combined light of sun and sky, for example, as reflected from the unshaded sheet of white paper illuminated outdoors. Significant color shifts occur in daylight, depending on geography, season and time of day, but it is unchanged by scattered clouds or if it is overcast: these only dim the light and diffuse it.

Some people face troubles when it comes to discerning colors, and the task of distinguishing different colors is extremely difficult. For these colorblind individuals, it is difficult when they try to color-coordinate their outfits or look at a stoplight [1, 5, 28].

When we took about imaging process and light spatial distribution, in addition to color vision, the human eyes can suffer from diseases and disorders [4, 9,24,36]. In recent years, the development of adaptive optics has allowed the quality of the images projected on the retina, as well as those taken of the retina, to be improved [11, 19, 27].

The human eye is an imperfect optical instrument, and there are not two individuals exactly alike, even their two eyes. Each individual has an optical eye print that is as unique as their own fingerprints. We distinguish one person's optical eye print from another depending on the eye imperfections or the eye aberrations. However, despite the fact that we can measure many orders of aberrations, the human eye can only distinguish aberrations up to the fourth or fifth order. Approximately 80% - 83% of visual aberration errors can be explained through the first and second-order aberrations. Lower order aberrations are typically diagnosed and measured by traditional refraction, automated or computerized refraction, keratometry and corneal topography. These techniques are quite economical from the financial point of view. The problem is that our imperfect vision from our imperfect eyes is typically corrected with perfect or standardized lenses that don't exactly match our individual optical eye prints [4,11,37].

Higher order aberrations are defined as any refractive error that cannot be corrected by spherocylindrical lenses combinations. To determinate the other higher order wave aberrations requires more sophisticated techniques. Wavefront measuring instruments like the Shack-Hartmann Sensor, different interferometer techniques or Talbot imaging are being utilized to provide data from which the wave aberrations of the human eye can be accurately and reliably estimated. Thesekinds of sensors can measure aberrations from three to five orders and are far more precise and expensive from the financial point of view. Currently, there is interest in correcting the higher-order optical aberrations of the human eye using laser eye surgery technique like LASIK (laser-assisted in situ keratomileusis). This technique corrects myopia, hypermetropia, defocus and astigmatism. Surgically correcting the additional higher-order aberrations could extend the limits of human vision beyond 20/20 vision-rating on the Snellen Acuity scale.

Alternatively, we can use a wavefront-guided LASIK, which is a variation of LASIK surgery in which, rather than applying a simple correction of focusing power to the cornea, as in traditional LASIK, an ophthalmologist applies a spatially varying correction, guiding the computer-controlled excimer laser with measurements from a wavefront sensor. In addition, this type of capability could be used to eliminate the higher-order aberrations that may have been caused by a prior surgical procedure.These induced aberrations are responsible for the glare, halos, and other visual defects [11,19,27,36]. This method's success depends a lot on the surgeon's skills.

If the patients want, they can use non-surgical options, which include eyeglasses and soft contact lenses or rigid gas permeable lenses.

2. Color Vision

Color vision is the capacity of an organism or machine to distinguish objects based on the wavelengths of the light they reflect or emit. The eye's cornea and lens focus the incident light into the light-sensitive receptors in our retina. The light stimulus incident at the cornea is transformed into an image at the retina.

The retinal image is transformed into a neural response by the light-sensitive elements of the eye, the photoreceptors. The photoreceptors' responses are transformed to a neural response on the optic nerve. The optic nerve representation is transformed into a cortical representation, and so forth. In function of the number and distribution of the cones photoreceptors in to the eyes, the colors vision can be divided in two categories: normal vision and colorblindness [5,15,24,35]. In order to perceive colors, our eyes are trying to describe a continuous light spectrum using just three numbers. Actually, the eyes make the conversion of light into RGB values. Human vision is capable ofdistinguishing a lot of different color hues by perceiving different variations of the countless different photon frequencies. This can be done by changing the color temperature of the illumination source. In this process, of importance are the cones photoreceptors. The cones are conventionally labeled according to the ordering of the wavelengths of the peaks of their spectral sensitivities: short (S), medium (M), and long (L) cone types, also sometimes referred to as blue, green, and red cones. Color blindness is a color vision deficiency and represents the inability to perceive differences between some of the colors that others can distinguish. It is most often of genetic nature but may also occur because of eye, nerve, or brain damage. Sometimes is called Daltonism after the name of the English chemist Jon Dalton, who published the first scientific paper related to this subject. There are three major types of color blindness; they are anomalous trichromacy, dichromacy, and monochromacy [5,15,35].

People with anamalous trichromacy have all three wavelength receptorsshort, medium, and longbut one of the three receptors has mutated and does not absorb light of that wavelength as efficiently as an un-mutated receptor. When the mutated receptor is the long wavelength (red) receptor, this condition is known as protanomaly. When the mutated receptor is the medium wavelength (green) receptor, this condition is known as deuteranomaly. Finally, when the mutated receptor is the short wavelength (blue) receptor, the condition is known as tritanomaly. People with dichromacy are missing an entire receptor type. When the missing receptor is the long wavelength (red) receptors, this condition is known as protanopia. For green cone missing, we have deuteranopia. Finally, for missing short blue cones, we have tritanopia. There is finally type of color blindness, monochromacy, which is the total inability to distinguish any colors. This condition, however, is extremely rare [5,15,35]. Color blindness affects approximately 8% of men and less than 1% of women. Protanopes, deuteranopes, and tritanopes are dichromats, and 2% of population is affected by these vision disabilities. They can match any color they see with some mixture of just two spectral lights (whereas normally, humans are trichromats and require three lights). These individuals normally know they have a color vision problem, and it can affect their lives on a daily basis. In the case of red-green colorblindness, the most common type of color deficiency, the individual has only cones that absorb red or green wavelengths but not both. Protanopes and deuteranopes see no perceptible differences between red and orange, respectively, yellow, and green [5,15,28,35].

2.1. The Spectral Power Distributions of the Visible Light

CIE (Commission Internationale de l'Eclairage) illuminants D were intended to represent various phases of daylight. They are specified by the following formulas. The relative spectral power distribution $S_D(\lambda)$, of a CIE daylight illuminant is defined by

$$S_D(\lambda) = S_0(\lambda) + M_1 S_1(\lambda) + M_2 S_2(\lambda) \tag{1}$$

where $S_0(\lambda)$, $S_1(\lambda)$, and $S_2(\lambda)$ are the mean and the most important "eigenvectors" of a large set of measured daylight distributions [12].

Figure 1 a) shows the spectral power distribution of the three CIE daylight vectors. M_1 and M_2 are scalar multipliers that are related to the chromaticity coordinates (x_D , y_D) of the illuminant by the following functions [12,31]

$$M_1 = \frac{-1.3515 - 1.7703x_D + 5.9114y_D}{0.0241 + 0.2562x_D - 0.7341y_D}, \quad (2)$$

$$M_2 = \frac{0.300 - 31.4424x_D + 30.0717y_D}{0.0241 + 0.2562x_D - 0.7341y_D}. \quad (3)$$

The chromaticity coordinates y_D is computed from x_D using relation

$$y_D = -3.000x_D^2 + 2.870x_D - 0.275 \quad (4)$$

which defines the "CIE daylight locus" in the CIE 1931 chromaticity diagram.

For correlated color temperature T_c between 4000 K to 7000 K

$$x_D = -4.6070\frac{10^9}{T_c^3} + 2.9678\frac{10^6}{T_c^2} + 0.9911\frac{10^3}{T_c} + 0.244063. \quad (5)$$

For correlated color temperature T_c between 7000 K to 25000 K

$$x_D = -2.0064\frac{10^9}{T_c^3} + 1.9081\frac{10^6}{T_c^2} + 0.24748\frac{10^3}{T_c} + 0.237040. \quad (6)$$

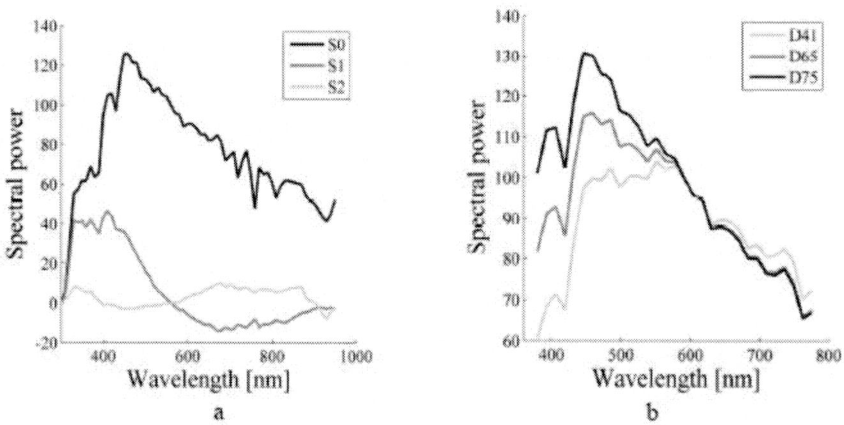

Figure 1. a)the spectral power distributions of the CIE daylight vectors $S_1(\lambda)$, $S_2(\lambda)$ and $S_3(\lambda)$, b) the spectral power distributions of the CIE daylights.

In order to reduce the number of illuminants used in colorimetry, CIE recommends that D65 $T_c = 6504K$ be used as the preferred illuminant. When it is not convenient to do so, D55 ($T_c = 5503K$) or D75 ($T_c = 7504K$) should be used [12]. Figure 1 b) shows the spectral power distributions of the CIE daylight illuminant at several correlated color temperatures.

2.2. Human Eye Responses to Color

The response of the L, M, S cones c_i, i = *S, L, M*, for spectral power distribution of the light $l(\lambda)$, on the λ luminous spectrum is

$$c_i = \int_{\lambda \min}^{\lambda \max} S_i(\lambda) l(\lambda) d\lambda \tag{7}$$

$S(\lambda)$ is the function of sensitivity of *i*-th type of cones, $[\lambda_{\min}, \lambda_{\max}]$ is an interval of visible spectrum, typically between 380 nm and 780 nm.

From mathematical point of view, equation (7) is a Fredholms equation of first order. If we define an integral operator as

$$Ll = \int_{\lambda \min}^{\lambda \max} s_i(\lambda) l(\lambda) d\lambda \tag{8}$$

the operator L is an operator thatprojects a Hilbert space into another Hilbert space; this Hilbert space is a space of square integrable functions [20]. We substituteequation (7) into equation (8), and we rewrite it in the form

$$c = Ll. \tag{9}$$

Equation (9) is a classic form of linear transformation where the spectrum of light is transformed into the cones responses through the inner product. This transformation is defined by the sensitivity functions S_i, i =L, M, S. Space thatis defined by the S_i,i =L, M, S, is called human visual subspace [32].

We rewrite the equation (7) in discrete form. Consequently, the color stimulus are no longer a continuous wavelength spectrum, instead we have N sample separated by a sampling step $\Delta\lambda$

$$\Delta\lambda = \lambda_i - \lambda_{i-1} = \frac{\lambda_{\max} - \lambda_{\min}}{N}, \tag{10}$$

$$c_i = \sum_{i=1}^{N} s_i(\lambda_i) l(\lambda_i) \Delta\lambda = s_i^T l \Delta\lambda \text{ i = S, L, M} \tag{11}$$

where $s_i = [s_i(\lambda_1), s_i(\lambda_2),...., s_i(\lambda_N)]^T$ and $l_i = [l(\lambda_1), l(\lambda_2),..., l(\lambda_N)]^T$,T denotes vector transpose operation. We rewrite equation (11) as a matrix

$$c = S_i^T l \Delta\lambda \quad (12)$$

where

$$c = [c_S, c_M, c_L]^T \ , s = \begin{bmatrix} s_{S1} & s_{M1} & s_{L1} \\ s_{S2} & s_{M2} & s_{L2} \\ \vdots & \vdots & \vdots \\ s_{SN} & s_{MN} & s_{LN} \end{bmatrix}.$$

From equation (12), we see that the color can be specified by the vector c[1x3],which is known as tristimulusvector.

If we perceive light that is reflected from a surface, instead of light that is directly emitted from a light source, our eyes receive result of the scalar product of reflectance and radiance spectrum. The mathematical expression of the human eye response to light is in continuous case is

$$c_i = \int_{\lambda \min}^{\lambda \max} S_i(\lambda) r(\lambda) l(\lambda) d\lambda \, i = S, L, M \quad (13)$$

where $r(\lambda)$ is the spectral reflectance of the surface.

For the discrete case, we have

$$c = S^T l r \Delta\lambda . \quad (14)$$

2.3. Metamerism

Colors l_1 and l_2 match if they produce the same visual response

$$S^T l_1 = S^T l_2 . \quad (15)$$

Metamerism happens if colors match from visual point of view; this fact does not mean that colors match from physical point of view, consequently these colors might have completely different spectral power distribution. This phenomenon is in the background of the color reproduction in CRT and other trichromatic monitors. Metamerism allows us to convert the system of coordinate (r, g, b) in to a new system of coordinate (x, y, z) and the curve of spectrally pure colors $\vec{r}, \vec{g}, \vec{b}$ (Figure 2a)) in to curve of spectrally colors as defined by $\vec{x}, \vec{y}, \vec{z}$ (Figure 2 b)) [15].

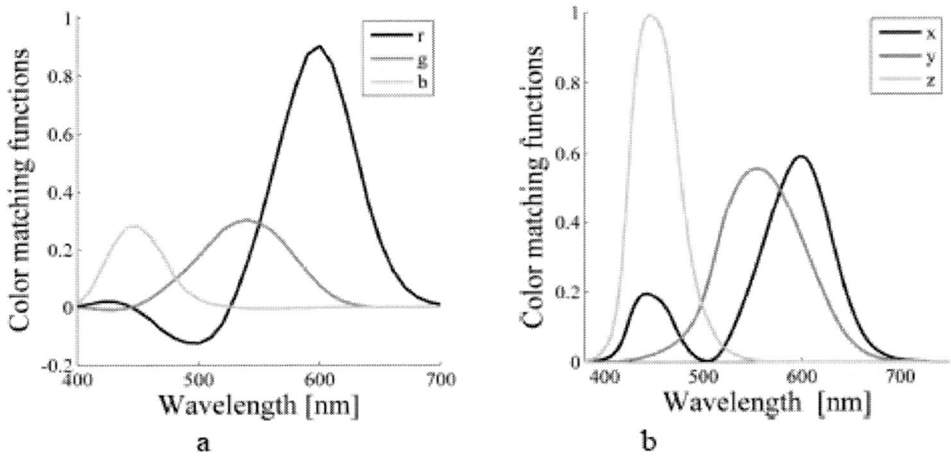

Figure 2.a) the color matching functions $\vec{r}(\lambda)$, $\vec{g}(\lambda)$, $\vec{b}(\lambda)$ for a 10 - deg. field (CIE 1964), b)the color matching functions $\vec{x}(\lambda)$, $\vec{y}(\lambda)$, $\vec{z}(\lambda)$, for a 10 - deg. field (CIE, 1964).

The chromaticity coordinates *r*, *g*, *b* are obtained by normalizing the tristimulus values

$$r = \frac{R}{R+G+B}, g = \frac{G}{R+G+B}, b = \frac{B}{R+G+B}, \tag{16}$$

$$r + g + b = 1, \tag{17}$$

$$R = \int_0^\infty P(\lambda)\vec{r}(\lambda)d\lambda, \tag{18}$$

$$G = \int_0^\infty P(\lambda)\vec{g}(\lambda)d\lambda, \tag{19}$$

$$B = \int_0^\infty P(\lambda)\vec{b}(\lambda)d\lambda \tag{20}$$

where $P(\lambda)$ is the spectral radiance,*r*, *g*, *b* are the chromaticity coordinates, $\vec{r}(\lambda)$, $\vec{g}(\lambda)$, $\vec{b}(\lambda)$ are the color matching functions,*R*, *G*, *B* are the tristimulus values.

The chromaticity coordinates *x*, *y*, *z* are obtained by normalizing the tristimulus values

$$x = \frac{X}{X+Y+Z}, y = \frac{Y}{X+Y+Z}, z = \frac{Z}{X+Y+Z}, \tag{21}$$

$$x+y+z = 1, \tag{22}$$

$$X = \int_0^\infty P(\lambda)\vec{x}(\lambda)d\lambda, \quad (23)$$

$$Y = \int_0^\infty P(\lambda)\vec{y}(\lambda)d\lambda, \quad (24)$$

$$Z = \int_0^\infty P(\lambda)\vec{z}(\lambda)d\lambda \quad (25)$$

where $P(\lambda)$is the spectral radiance,x, y, z are the chromaticity coordinate, $\vec{x}(\lambda)$, $\vec{y}(\lambda)$, $\vec{z}(\lambda)$ are the color matching functions,X, Y, Z are the tristimulus values.

2.4. The X Y Z Color Space

In 1931, CIE (Committee Internationale d'Eclairage) decided to propose a particular set of color matching functions as a standard. These functions are denoted as $\vec{x}(\lambda)$, $\vec{y}(\lambda)$, $\vec{z}(\lambda)$, (shown in Figure 2 b)). According to equation (11), the corresponding tristimulus values X, Y, Z for stimuli f are

$$\begin{aligned} X &= k\sum_{\lambda}\vec{x}(\lambda)f(\lambda)\Delta\lambda \\ Y &= k\sum_{\lambda}\vec{y}(\lambda)f(\lambda)\Delta\lambda \\ Z &= k\sum_{\lambda}\vec{z}(\lambda)f(\lambda)\Delta\lambda \end{aligned} \quad (26)$$

where k is a constant normalization factor.

We can rewrite (26) in compact form by using the notation as in equation (12)

$$S = [\vec{x}\vec{y}\vec{z}]$$

$$S = \begin{bmatrix} x_1 & y_1 & z_1 \\ x_2 & y_2 & z_2 \\ \vdots & \vdots & \vdots \\ x_N & y_N & y_N \end{bmatrix}. \quad (27)$$

Here, each row is intensities to match a single monochromatic light, and each column is a complete set of intensities over wavelength (with amount of samples N) of one primary.

Consequently, equation (26) becomes

$$f_{xyz} = kS^T f \Delta\lambda \tag{28}$$

where

$$k = \frac{100}{\sum_{\lambda} l_{\lambda} \bar{y}_{\lambda} \Delta\lambda}.$$

The relation between the XYZ and RGBis

$$\begin{bmatrix} X \\ Y \\ Z \end{bmatrix} = \begin{bmatrix} 0.4125 & 0.3576 & 0.1804 \\ 0.2127 & 0.7152 & 0.0722 \\ 0.0193 & 0.1192 & 0.9503 \end{bmatrix} \cdot \begin{bmatrix} R \\ G \\ B \end{bmatrix}. \tag{29}$$

2.5. The ITU-R BT.709 Standard

The ITU-R BT.709 standard specifically describes the encoding transfer function for a video camera that when viewed on a standard monitor will produce excellent image quality. This recommendation specifies the encoding of real world scene tristimulus values into a standard monitor RGB color space assuming a dark viewing condition. A common choice of primaries for a display device is [6,18,20,23,26,30]

$$M_{709} = \begin{bmatrix} 0.640 & 0.300 & 0.150 \\ 0.330 & 0.600 & 0.060 \\ 0.030 & 0.100 & 0.790 \end{bmatrix}. \tag{30}$$

The white pointis the set of tristimulus values of the color reproduced by equal contributions of the red, green, and blue primaries. White point serves to define the color white in image capture or reproduction. A commonly used white point is the D_{65} white point

$$x_{D65} = 0.3127\,;\ y_{D65} = 0.3290\,;\ z_{D65} = 0.3582. \tag{31}$$

Gamma represents a numerical parameter that describes the nonlinearity of luminance reproduction. Phosphors of monitors do not react linearly with the intensity of the electron beam. Instead, the input value is effectively raised to an exponent called gamma.The gamma function is nonlinear [2,6]. The purpose of the gamma function is to increase the luminosity of the image that is rendered on the diplay, and is a correction done by the hardware of the monitor.Gamma correction is, in the simplest cases, defined by the following power law expression

$$V_{out} = V_{in}^{\gamma}. \tag{32}$$

2.6. The Color Blindness

Normal color vision requires the use of L, M, S cones, which are located in the retina of the eye. An abnormality, or deficiency, of any of the types of cones will result in abnormal color vision. The colors blindness can be divided in three types: anomalous tichromacy, dicromacy and monocromcay [5,15,35].

The anomalous trichromacy is a mild shift in the sensitivity of pigments of the cones. It can be divided in three types: protanomaly, deuteranomaly and tritanomaly. In the protanomaly, the shades of red appear weaker in depth and brightness.

In the deuteranomaly, the shades of green appear weaker in depth and brightness. The tritanomaly is very rare case where shades of blue appear weaker in depth and brightness.

The dichromacy is a great deficiency or missing completely of one of the cones. It can be of three types: protanopia, deuteranopia and tritanopia. In the protanopia, the shades of red are greatly reduced. In the deuteranopia, the shades of green are greatly reduced. The tritanopia is a very rare case where shades of blue are greatly reduced.

Achromatopsia means "without color" and is defined as little or no function of the cone cells. There are two primary forms, the complete achromatopsia and incomplete achromatopsia. Persons with achromatopsia are only able to perceive black, white and shades of gray. Patients with complete achromatopsia have no real understanding of the concept of color. Patients with incomplete achromatopsia have a small residual amount of color vision due to the presence of some functioning cone cells in the retina [5,15,35].

In order to simulate the dicromaticy vision, we make an algorithm that converts the RGB (red, green, blue) normal viewer image to LMS (long, medium, short).

The LMS values correspond to what is received at the level of the retina. RGB, on the other hand, corresponds to the phosphor levels on a cathode ray tube screen to match the colors and is used to define digital images. We focus our attention onthe protanopia and deuteranopia, the most frequent dicromaticy situation

$$\begin{bmatrix} L \\ M \\ S \end{bmatrix} = \begin{bmatrix} 17.8824 & 43.5161 & 4.11935 \\ 3.45565 & 27.1554 & 3.86714 \\ 0.0299566 & 0.184309 & 1.46709 \end{bmatrix} \cdot \begin{bmatrix} R \\ G \\ B \end{bmatrix}. \tag{33}$$

In the next step, we make a conversion to delete the information associated with the loss of any of the cone types to get the modified LMS values L'M'S'. The matrix formulation for the protanopia and deuteranopia

$$\begin{bmatrix} L_p \\ M_p \\ S_p \end{bmatrix} = \begin{bmatrix} 0 & 2.02344 & -2.52581 \\ 0 & 1 & 0 \\ 0 & 0 & 1 \end{bmatrix} \cdot \begin{bmatrix} L \\ M \\ S \end{bmatrix}, \tag{34}$$

$$\begin{bmatrix} L_d \\ M_d \\ S_d \end{bmatrix} = \begin{bmatrix} 1 & 0 & 0 \\ 0.494207 & 0 & 1.24827 \\ 0 & 0 & 1 \end{bmatrix} \cdot \begin{bmatrix} L \\ M \\ S \end{bmatrix}. \tag{35}$$

Finally, we make a reverse transformation on the L'M'S' values to get the R'G'B' vales. R'G'B' presumably represent how that specific color RGB is perceived by a color blind person [1,28,34]

$$\begin{bmatrix} R \\ G \\ B \end{bmatrix} = \begin{bmatrix} 0.0809 & -0.1305 & 0.1167 \\ -0.0102 & 0.0540 & 0.1136 \\ -0.00036 & -0.0041 & 0.6935 \end{bmatrix} \cdot \begin{bmatrix} L \\ M \\ S \end{bmatrix}. \quad (36)$$

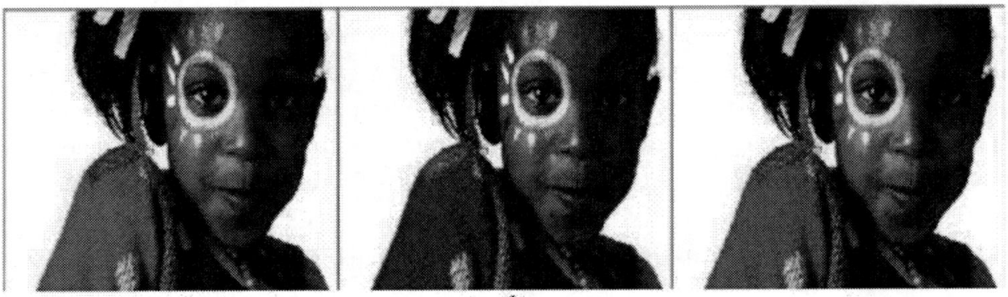

a b c

Figure 3.The D55 illumination, a) the normal vision, b) the deuteranopia, c) the protanopia.

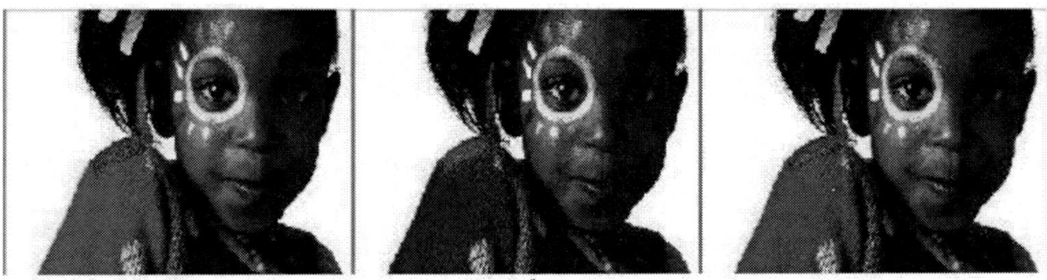

a b c

Figure 4. The D75 illumination, a) the normal vision, b) the deuteranopia, c) the protanopia

From Figure 3 and Figure 4, we see that the normal viewer and the color blindness viewer are both sensitive to color temperature; consequently, the viewers can distinguish different hues of the scene colors.

3. The Eye Acuity

The eye is a sense organ that acts as a photoreceptor capable of image formation. The eye is an imperfect optical instrument, and there are no two individuals exactly alike, even their two eyes. Each individual has an optical eye print that is as unique as their own fingerprints. Optometry is the science that measures the visual acuity of the eye in order to determine the refractive errors of the eyes. Based on these measurements, an ophthalmologist can prescribe and fit appropriate corrective lenses [4,9,24] or recommend eye surgical procedures that correct the acuity of the eye and improve the resolution of the vision.

Visual acuity represents the sharpness of vision, especially as tested with a Snellen chart. Normal visual acuity based on the Snellen chart is 20/20. Visual acuity is the ability to distinguish details. Acuity may be specified in terms of object size, retinal size, or in angular or inverse angular units. There are four categories of acuity: detection, recognition, resolution, and vernier [4,9,24].

The detection of a bright sub-resolution object, such as a star, depends only on its flux and is not a matter of acuity.

Recognition is measured by the familiar Snellen eye chart. A chart for testing visual acuity, usually consisting of letters, numbers, or pictures printed in lines of decreasing size, which a patient is asked to read or identify at a fixed distance.

Resolution is the ability to perceive the separation of patterns that would be visible if present singly.

Vernier acuity is the ability of the eye to align objects relative to one another.

3.1. The Resolution of the Optical System

When we work with optical components, the most important problem is that it is impossible to image a point object as a perfect point image. An optical system is made by a set of components (surfaces) through which the light passes. The optical sensor is analyzed in space by the PSF (point spread function) and in the spatial frequency by the MTF (modulation transfer function), which are the most important integrative criterion of imaging evaluation for the optical system [2,3,13].

The PSF gives the 2D intensity distribution about the image of a point source. PSF gives the physically correct light distribution in the image plane including the effects of aberrations and diffraction. The errors are introduced by design (geometrical aberrations), optical and mechanical fabrication or alignment. MTF characterize the optical system functionality in spatial frequencies.

Most optical systems are expected to perform a predetermined level of image integrity. MTF describes the image structure as a function of spatial frequency and is specified in lines per millimeter. It is obtain by the Fourier transform in the image spatial distribution or spread function.

When an optical system processes an image using incoherent light, then the function which describes the intensity in the image plane produced by a point in the object plane is called the impulse response function [2,3,13]

$$g(x,y)=H\left\lfloor f(x,y)\right\rfloor \tag{37}$$

where H is an operator representing a linear, position (or space) invariant system.

The pattern of the input object intensity and the pattern of the output image intensity are related by a simple convolution equation

$$g(x,y)=\int\int_{-\infty}^{+\infty} f(\alpha,\beta)H\left[\delta\left((x-\alpha,y-\beta)\right)\right]d\alpha d\beta$$

$$g\ x,y\ = \int\int_{-\infty}^{+\infty} f\ \alpha,\beta\ h\ x-\alpha,y-\beta\ d\alpha d\beta \tag{38}$$

where in optics, it is called PSF (the point spread function)[2,3,13].

The PSF characterizes the image analyses in space, but we can also characterize the image in frequency using the OTF(optical transfer function)

$$OTF\ \alpha,\beta\ = \frac{FT\ PSF}{FT\ PSF\ \big|_{\alpha=0,\beta=0}}. \tag{39}$$

The MTF is a quantitative measure of the ability of an optical system to transfer various levels of detail from object to image. The MTF (modulation transfer function) is defined as: the ratio of the contrast of the output image to that of the input image

$$MTF\ = \frac{contrast\ of\ the\ output\ image}{contrast\ of\ the\ input\ image}. \tag{40}$$

The OTF describes the response of the optical system to a known input and the relation between *OTF* and MTF is

$$MTF = |OTF|. \tag{41}$$

The optical resolution is a measure of the optical instruments' ability to resolve the details present in the scene. In optics, spatial resolution is expressed as contrast or MTF. An optical system is typically judged by its resolving power or ability to reproduce fine detail in an image. One criterion, attributed to Lord Rayleigh, is that two image points are resolved if the central diffraction maximum of one is no closer than the first diffraction zero of the other. Rayleigh's criterion, applied to point images, demands that the images be separated by 0.61distance between centers of the dots [3,13].

In Figure 5, we see three possible situations, corresponding to separations of 0.4, 0.5, and 0.6. The first case is not resolved because there is no evidence between the two peaks. The second case is barely resolved, and the third case is adequately resolved.

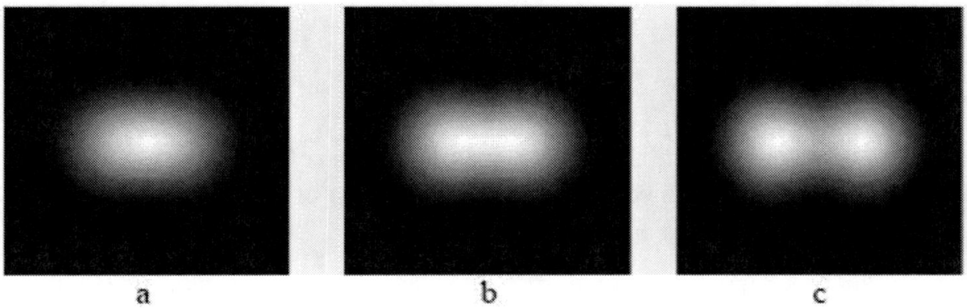

Figure 5. The Rayleigh's criterion a) 0.4 separation, c) 0.5 separation, d) 0.6 separation.

4. The Aberrations of the Eye

The wavefront of the light that is transmitted through an optical system is an imaginary surface that remains normal to the direction of propagation at all cross-sectional points within the optical pathway. For a perfect eye focused at infinity, the wavefront of the light collected by the optics of the eye would be part of an aspheric surface, which would converge on the back of the eye to create a (diffraction-limited) spot on the retina. Because of the reciprocal behavior of light (i.e., it traverses the same path in either direction), it is possible to measure this wavefront from light that is scattered or reflected from the retina. The light scatters from the retinal surface; it is collected by the lens and cornea and is then projected out of the eye. A perfect, emmetropic eye would completely collimate this light. The resulting wavefront would be a perfect flat plane wavefront, perpendicular to the direction of propagation. Any deviations from a perfect plane wave are determined by the optical errors in the eye, and as a result are called the wavefront error of the system [9,11,19,27,30].

To determine and apply the proper correction that will eliminate or minimize aberrations, their characteristics must first be captured and quantitatively described. The wave aberration function suits this purpose well because it completely describes the aggregate effects of the cornea and lens on light passing through every location in the pupil. It defines how the phase of the light is affected after it has propagated through the optical system. In the case of laser eye surgery, if we know the wave aberration function, we know what must be done to the refractive properties of the eye in order to create an image on the retina that is limited only by diffraction and scattering.

4.1. The PSF with Aberrations

The image of a point object formed by the optical system is given by the point spread function or impulse response. It is defined as [3,13]

$$PSF(x,y) = \frac{1}{\lambda^2 d^2 A_p} \left| F\left[p(x,y) \cdot e^{-i\frac{2x}{\lambda}W(x,y)} \right] \right|^2 \tag{42}$$

where$F$ is the Fourier transform operator,d is the distance from the exit pupil to the image, A_p is the aria of the exit pupil, λ is the wavelength, $e^{-i\frac{2x}{\lambda}W(x,y)}$ is the phase deviation of the wavefront from a reference sphere,$W(x,y)$ is the wave aberration function at the exit pupil

$$P(x,y) = p(x,y) \cdot e^{-i\frac{2x}{\lambda}W(x,y)} \tag{43}$$

where$P(x, y)$ is the generalized exit function.

The effect of aberrations on the OTF is to reduce the contrast of the image. There is no combination of optical aberrations that will increase contrast at any spatial frequency above the diffraction limit at that frequency.

4.2. The Monochromatic Siedel Aberrations

Aberrations are the failure of the light rays emerging from a point object to form a perfect point image after passing through an optical system. Aberrations lead to blurring of the image, which are produced by the image-forming optical system. To describe the primary monochromatic aberrations of rotationally symmetrical optical systems, we specify the shape of the wavefront emerging from the exit pupil. For each object point, there will be a quasi-spherical wavefront converging toward the paraxial image point.

In Figure 6, the wave aberration function $W(r,\theta)$ is defined as the distance, in optical path length, from the reference sphere to the wavefront in the exit pupil measured along the ray as a function of the transverse coordinates (x, y) of the ray intersection with a reference sphere centered on the ideal image point [9,11,13,38,39].

The aberrations are given by the Siedel's formula:

$$W(r,\theta) = W_{020}r^2 + W_{040}r^4 + W_{131}hr^3\cos(\theta) + W_{220}h^2r^2 + W_{311}h^3r\cos(\theta) + (higher\quad order\quad terms) \tag{44}$$

where W_{klm} are the coefficients of the wave aberration for the various terms of modes, h is the height of the object, r,θ are the polar coordinates in the pupil plane, r^2 Defocus, r^4 Spherical Aberration, $hr^3\cos(\theta)$ Coma, $h^2r^2\cos^2(\theta)$ Astigmatism, h^2r^2 Field Curvature, $h^3r\cos(\theta)$ Distortion.

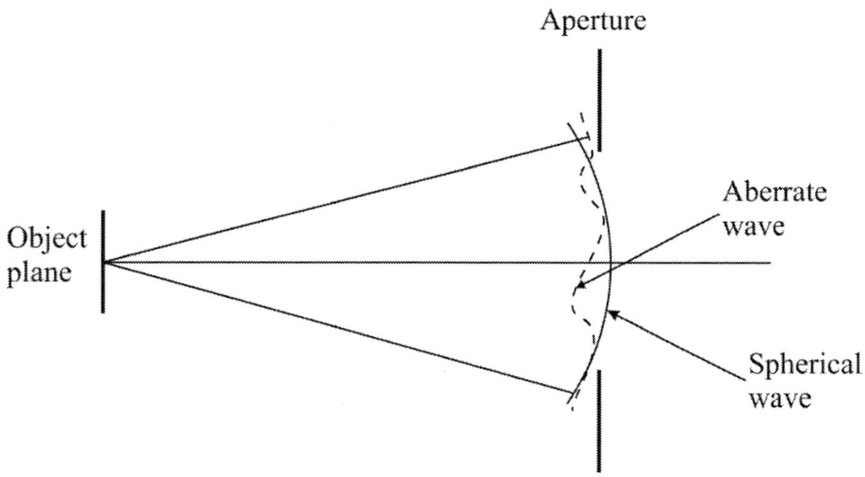

Figure 6. The wavefront aberration.

The Seidel's aberrations formula represents orthogonal polynomials, which have the next properties: field aberrations describe the wavefront for a single object point as a function of pupil coordinates (r,θ) and field height h. The aberrations are described functionally as a linear combination of polynomials. Point aberrations depend only on pupil coordinates, and each polynomial term represents a single aberration. The aberration polynomial may be extended to higher order; these are all the terms to fourth order. Ray aberrations are described by a polynomial one order lower than the corresponding wave aberrations [13,14,27,30,38,39].

4.3. The Zernike Polynomials

Zernike polynomials are recommended to describe wavefront aberrations from the mathematical point of view. The Zernike polynomials form a complete set of functions or modes that are orthogonal over a circle of unit radius and are convenient for serving as a set of basic functions.This makes them suitable for accurately describing wave aberrations as well as for data fitting.

They are usually expressed in polar coordinates and are readily convertible into Cartesian coordinates.These polynomials are mutually orthogonal and are therefore mathematically independent, making the variance of the sum of modes equal to the sum of the variances of each individual mode.They can be scaled so that non-zero order modes have zero mean and unit variance. [9,11,19,24,39].

The Zernike polynomials are defined as

$$Z_n^m(\rho,\theta) = N_n^m R_n^{|m|}(\rho)\cos(\theta,m) \quad for \quad m \geq 0, \quad 0 \leq \rho \leq 1 \quad 0 \leq \rho \leq 2\pi, \tag{45}$$

$$Z_n^m(\rho,\theta) = -N_n^m R_n^{|m|}\sin(\theta,m) \quad for \quad m < 0 \quad 0 \leq \rho \leq 1 \quad 0 \leq \rho \leq 2\pi \tag{46}$$

wherefor a given n; m can only take on values of $-n$, $-n+2$, $-n+4$,...n, N_n^m is the normalization factor, $R_n^m(\rho)$ is the radial polynomial,

$$N_n^m = \sqrt{\frac{2(n+1)}{1+\delta_{m0}}} \quad \delta_{m0} = 1 \text{ for} m\text{=0}, \ \delta_{m0} = 0 \text{ for } m \neq 0$$

$$R_n^{|m|}(\rho) = \frac{(-1)(n-s)!}{s!\left[0.5(n+|m|)-s\right]!\left[0.5(n-|m|)-s\right]!}\rho^{n-2s} \tag{47}$$

In Table 1, we have a list with the first ten Zernike polynomials.

In Figure 7, we have the graphical representation of the Zernike polynomials of order two, and in Figure 8 we have the PSF of the polynomials from Figure 7.

The wave aberration function can be express as a weighted sum of Zernike polynomials

$$W(\rho,\theta)=\sum_{n}^{k}\sum_{m=-n}^{n}W_n^m Z_n^m(\rho,\theta)=$$

$$\sum_{n}^{k}\left\{\sum_{m=-n}^{-1}\left[-N_n^m R_n^{|m|}(\rho)\sin(m\theta)\right]+\sum_{m=0}^{n}\left[N_n^m R_n^{|m|}(\rho)\cos(m\theta)\right]\right\} \tag{48}$$

where k is the polynomial order of the expansion, W_n^m is the coefficient of the Z_n^m mode in expansion and is equal to rms wave front error for that mode.

We can express the Zernike Polynomials in Cartesian coordinate

$$W(x,y)=\sum_{j=0}^{j_{\max}}W_j Z_j(x,y) \tag{49}$$

where

$$W_j=W_n^m,\ Z_j=Z_n^m(x,y),\ j\frac{n(n+2)+m}{2}.$$

Table 1. The first ten Zernike polynomials

N	M	N	Polynomials	Name
0	0	0	1	Piston
1	1	1	$2\rho\cos(\theta)$	Tilt x
	-1	2	$2\rho\sin(\theta)$	Tilt y
	0	3	$\sqrt{3}(2\rho^2-1)$	Power
2	2	4	$\sqrt{6}\rho^2\cos(2\theta)$	Astigmatism x
	-2	5	$\sqrt{6}\rho^2\sin(2\theta)$	Astigmatism y
	1	6	$2\sqrt{2}(3\rho^2-2)\rho\cos(\theta)$	Coma x
	-1	7	$2\sqrt{2}(3\rho^2-2)\rho\sin(\theta)$	Coma y
	0	8	$\sqrt{5}(6\rho^4-6\rho^2+1)$	Primary Spherical
3	3	9	$\rho^3\cos(3\theta)$	Trefoil x

a b c

Figure 7. The Zernike polynomials of order two, a) primary spherical,b) coma x,c) astigmatism x.

a b c

Figure 8. The PSF of the Zernike polynomials of order two, a) primary spherical, b) coma x,c) astigmatism x.

5. The Sources of Blur in Retinal Image

In this paragraph, we investigate two sources of blur in retinal images. These sources are the diffraction and the aberrations of the eye. We use principles of Fourier optics in order to characterize their effects on human vision [3,9,11,23]. The aberrations of the eye are simulated using the Zernike polynomials.

5.1. The Effect of Diffraction on Human Vision

Even for a perfect optical eye, the quality of the image on the retina is not perfect because it is deteriorated by the diffraction that happens when the light enters into the eye trough the pupil.

Diffraction means that the light spreads whenever it passes through the pupil. Thus, diffraction is the ultimate limit on image quality in any optical system. The degree of spreading is greater for smaller pupils and shorter wavelengths of light (Figure 9). In Fourier optics, the aperture's functionality is characterized by the point spread function.

The pupil of the eye has a circular shape [3]

$$c(r) = circ\left(\frac{r_0}{r}\right) \tag{50}$$

where r is the radius of the circle, r_0 is the cut off radius.

The PSF is calculated as

$$c(x,y) = \lambda \frac{r_0}{r} J_1\left(\lambda \frac{r_0}{r}\right) \tag{51}$$

where λ is the wavelength, J_1 is the Bessel function of order one.

A perfect optical system is diffracted limited by relation

$$d = 2.44\lambda N \tag{52}$$

where

$$N = \frac{f}{d} \tag{53}$$

where f is the focus length, d is the diameter of the aperture.

a b

Figure 9. a) the diffraction of 8 mm pupil, b) the diffraction of 3 mm pupil.

5.2. The Effect of the Aberrations on Human Vision

In any real eye, optical performance is worse than predicted by diffraction because of imperfections in the optics. Aberrations obscure the Airy disc pattern due to diffraction and produce a larger and more complex light distribution.

In general, aberrations affect the light rays that enter the edge of the pupil more strongly than they affect rays entering the center of the pupil. In general, diameter of the pupil is between 3 mm and 8 mm.

In Figure 10, we simulate the effects of the aberrations presented in Paragraph 3.3 for a pupil of diameter 8 mm [9, 11, 23].

a b c

Figure 10.The effect of the aberrations on human vision a) primary spherical, b) coma x, c) astigmatism x.

6. The Measurement of the Wavefront Errors

In this chapter, we present three different techniques for measurement of the aberrations of the eye: the Shark Hartman sensor, interferograms and the Talbot imaging.

When are applied to the human eye, the wavefront sensing reveals aberrations that aren't detectable by conventional means. The ocular wavefront can be distorted by imperfections in any part of the visual system, including the cornea, lens and vitreous.

The Shack-Hartmann wavefront sensor measures the shape of the wavefronts of light that exit from the pupil of the eye. If the eye is a perfect optical system, these wavefronts would be perfectly flat. Since the eye is not perfect, the wavefronts are not flat and have irregular curved shapes.

An interferometer can measure small wavefront deformations withhigh accuracy of the order of a fraction of the wavelength. The two-wave interferometers produce an interferogram by superimposing two wavefronts, one of which is typically a flatreference wavefront, and the other a distorted wavefront whose shape is to be measured.

Talbot imaging is a diffraction phenomenon that occurs with periodic objects. Perfect replicas of the object appear at fixed distances called Talbot planes. The location of the Talbot planes depends on the period of the object and the wavelength of light. If aberrations are present, the Talbot images are distorted. The amount of distortion depends upon the wavefront slope in the direction perpendicular to the bars of the object.

6.1. The Shark Hartman Sensor

A planar wavefront entering the eye form a small focused spot of light on the retina that is limited only by the effects of diffraction. The reflected light propagates back through the optics of the eye, and a wavefront sensor can measure the exiting light. Knowing the wavefront aberrations, we can determine the corrections that need to be made to the eye. The eye function is correct if an existing planar wavefront is formed from a point source on the retina; otherwise we have aberrations [10,11,14,16,19,24,25,36].

In Figure 11, a polarized light source is used to produce a small spot on the retina. The diffuse reflected light that propagates back through the lens and cornea and out of the eye is relayed to the plane of the lenslet array.

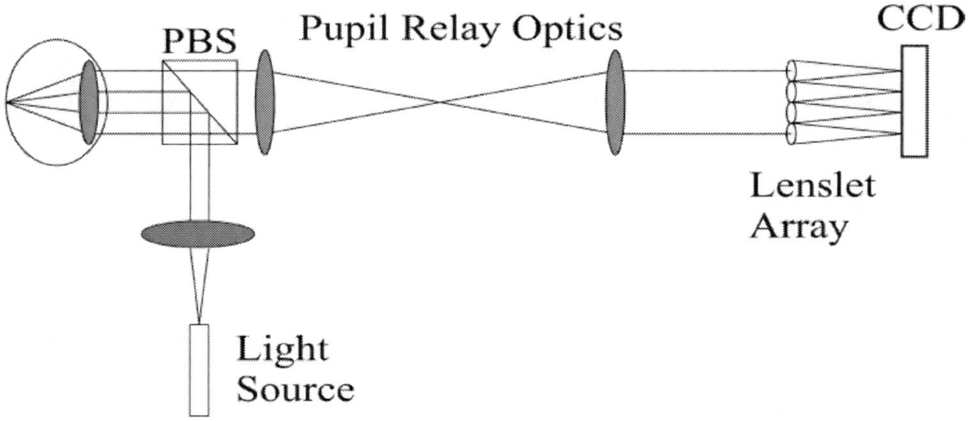

Figure 11. The schematic of the Shack-Hartmann wavefront sensor.

Any specularly reflected light will be reflected out of the return path by the polarization beam splitter (PBS). The Shack-Hartmann sensor uses a lenslet array to divide the subject target pupil into a field of smaller, non-overlapping "sub-pupils" for analysis. The lenslet array will convert the local sections of the wavefront into focused spots of the CCD.

If the incident wavefront is a perfect plane, then the focused spots will not be displaced from the optical axes of the various lenslets. An aberration incident wavefront will produce a non-zero spot displacement. For each lenslet, there will be a x-displacement, Δx, and a y-displacement, Δy [11,14,19,27]. The displacements divided by the focal length of the lensletareequal to the local slope of the wavefront. The displacement data measured by the CCD can now be fit to a Zernike polynomial expansion, where the expansion coefficients are determined using least-squares estimation

$$\frac{\partial W(x,y)}{\partial x} = \frac{\Delta x(x,y)}{f},$$

$$\frac{\partial W(x,y)}{\partial y} = \frac{\Delta y(x,y)}{f},$$

$$W(x,y) = \sum_j W_j Z_j(x,y)$$

W_j is the coefficient of the Z_j mode and is equal to the rms wavefront error for that mode

$$\frac{\partial W(x,y)}{\partial x} = \sum_j W_j \frac{\Delta Z_j(x,y)}{\partial x},$$

$$\frac{\partial W(x,y)}{\partial x} = \sum_j W_j \frac{\Delta Z_j(x,y)}{\partial x},$$

$$\frac{\Delta x(x,y)}{f} = \sum_j W_j \frac{\Delta Z_j(x,y)}{\partial x}, \tag{54}$$

$$\frac{\Delta y(x,y)}{f} = \sum_j W_j \frac{\Delta Z_j(x,y)}{\partial y} \tag{55}$$

Let

$$\frac{\Delta x(x,y)}{f} = b(x,y) \text{ and } \frac{\Delta y(x,y)}{f} = c(x,y),$$

$$\frac{\Delta Z_j(x,y)}{\partial x} = g_j(x,y) \text{ and } \frac{\Delta Z_j(x,y)}{\partial x} = g_j(x,y).$$

The equations (54) and (55) can be expressed in the matrix form

$$\begin{bmatrix} b(x_1,y_1) \\ b(x_2,y_2) \\ \vdots \\ b(x_k,y_k) \\ c(x_1,y_1) \\ c(x_2,y_2) \\ \vdots \\ c(x_k,y_k) \end{bmatrix} = \begin{vmatrix} g_1(x_1,y_1) & g_2(x_1,y_1) & \cdots & g_{j\max}(x_1,y_1) \\ g_1(x_2,y_2) & g_2(x_2,y_2) & \cdots & g_{j\max}(x_2,y_2) \\ \vdots & \vdots & \cdots & \\ g_1(x_k,y_k) & g_2(x_k,y_k) & \cdots & g_{j\max}(x_k,y_k) \\ h_1(x_1,y_1) & h_2(x_1,y_1) & \cdots & h_{j\max}(x_1,y_1) \\ h_1(x_2,y_2) & h_2(x_2,y_2) & \cdots & h_{j\max}(x_2,y_2) \\ \vdots & \vdots & \cdots & \vdots \\ h_1(x_k,y_k) & h_2(x_k,y_k) & \cdots & h_{j\max}(x_k,y_k) \end{vmatrix} \cdot \begin{bmatrix} W_1 \\ W_2 \\ \vdots \\ W_k \end{bmatrix} \quad 2k^2 > j_{\max} \tag{56}$$

With the matrix form, we can determine the Zernike polynomials using least-squares estimation. The reconstruction wavefront $\tilde{W}(x,y)$ is given by a linear superposition of the fitting polynomials

$$\tilde{W}(x,y) = \sum_{j=0}^{j_{\max}} \tilde{W}_j \tilde{Z}_j(x,y). \tag{57}$$

To determine the “goodness of the fit,” we can compare the derivative of the reconstructed wavefront with the derivatives of the originally wavefront

$$\text{rms error} = \sqrt{\sum_{i=1}^{N}\left[\left(\frac{\partial W_i}{\partial x} - \frac{\partial \tilde{W}_i}{\partial x}\right)\right] + \left(\frac{\partial W_i}{\partial y} - \frac{\partial \tilde{W}_i}{\partial y}\right)} \tag{58}$$

where the subscripts i denotes evaluation at the point (x_i, y_i).

The Shack Hartman wavefront sensors could measure the aberrations in clinical conditions such as dry eyes, keratoconus, post-laser insitukeratomileusis and LASIK surgery. The disadvantage of the Shack Hartman wavefront sensor is that in the eyes with a large amount of aberrations or extreme irregular corneas, crossover can occur. The crossover creates a false measurement or aninability of the device to calculate the aberration.

6.2. The InterferogramWavefront Aberrations Measurement

Interference is the superposition and composition of two or more coherent waves with constant phase difference. Interference can be constructive or destructive. We analyze the case with interference of two waves

$$I = I_1 + I_2 + 2\sqrt{I_1 I_2}\cos(\alpha_1 - \alpha_2) \tag{59}$$

where $\alpha_1 - \alpha_2$ is the phase difference between the two interfering beams, $\alpha_1 - \alpha_2 = \frac{2\pi}{\lambda}$ is the optical path difference.

Two-wave interferometers produce an interferogram by superimposing two wavefronts, one of which is typically a flat reference wavefront and the other a distorted wavefront whose shape is to be measured [14,39].

An interferometer can measure small wavefront deformation with a high accuracy of the order of a fraction of the wavelength.

Let us consider a two-wave interferometer with a flat wavefront that has a positive tilt about the y-axis and a wavefront under analysis, for which the deformations with respect to a flat wavefront without tilt are given by $W(x,y)$

$$I(x,y) = I_1(x,y)e^{ikW(x,y)} + I_2(x,y)e^{ikx\sin\theta} \tag{60}$$

where I_1 is the amplitude of the light beam at the wavefront under analysis, and I_2 is the amplitude of the light beam with the reference wavefront, and $k = 2\pi/\lambda$. The irradiance is

$$I(x,y) = I_1(x,y) + I_2(x,y) + 2\sqrt{I_1(x,y)I_2(x,y)}\cos k[x\sin\theta - W(x,y)] \tag{61}$$

where $I_1(x,y)$ and $I_2(x,y)$ are the irradiances of the two beams, and the phase difference between them is given by $\phi = k[x\sin\theta - W(x,y)]$.

In this chapter, we focus our attention on interferograms of the aberrations. We can generalize the analysis presented in equation (60) by summing more waves with different aberrations.

For example, in Figure 12, an eye can suffer from multiple aberrations, in our case: astigmatism x, coma x and primary spherical. The final aberration is the composition between all three aberrations [13,14,27].

Figure 12.The interferograms of themonochromatic aberrations a) astigmatism x,b) astigmatism x + coma x,c) astigmatism x + coma x + primary spherical.

6.3. The Talbot Imaging

The Talbot effect is a self-imaging phenomenon where the diffraction pattern from a periodic structure reforms a copy of the structure at downstream planes.

The effect can be derived from the Fresnel diffraction integral. We suppose that we have a periodic structure running parallel to the axis z from the plane (x,y) to the plane (x_1,y_1), then the electrical field can be represented as [3,36]

$$E(x_1,y_1) = \frac{\exp(ikz)}{i\lambda z}\exp\left[\frac{i\pi}{\lambda z}\left(x_1^2+y_1^2\right)\right]F\left\{E(x,y)\exp\left[\frac{i\pi}{\lambda z}\left(x^2+y^2\right)\right]\right\} \tag{62}$$

In order to simplify our calculus, we note

$$A = \frac{\exp(ikz)}{i\lambda z}\exp\left[\frac{i\pi}{\lambda z}\left(x_1^2+y_1^2\right)\right], \tag{63}$$

$E(x,y)$is a periodic function with period L, which can be expressed as Fourier series

$$E(x,y)=\sum_{n=-\infty}^{\infty} a_n \exp\left[\frac{i2\pi nx}{L}\right], \tag{64}$$

$$a_n=\frac{1}{L}\int_{=L/2}^{L/2} E(x,y)\exp\left[-\frac{i2\pi nx}{L}\right]. \tag{65}$$

We substituteequation (64) into equation (62)

$$E(x_1,y_1)=A\cdot F\left\{a_n\sum_{n=-\infty}^{\infty}\exp\left[\frac{i2\pi nx}{L}\right]\exp\left[\frac{i\pi}{\lambda z}\left(x^2+y^2\right)\right]\right\}. \tag{66}$$

Let $b^2 = i\lambda z$, and we use the properties that the Fourier transform of the product is equal to the convolution of the individual Fourier transforms

$$E(x_1,y_1)=A\cdot\sum a_n\left\{F\left[\exp\left(\frac{i2\pi nx}{L}\right)\right]\right\}*F\left[\exp\left(\frac{-\pi}{b^2}\left(x^2+y^2\right)\right)\right], \tag{67}$$

$$E(x_1,y_1)=b^2\cdot A\cdot\sum a_n\left[\delta\left(\xi-\frac{n}{L}\right)\delta(\eta)*\exp\left[-\pi b^2\left(\zeta^2+\eta^2\right)\right]\right], \tag{68}$$

F is the Fourier transform withrespect to variable $\xi=\frac{x_1}{\lambda z}$ and $\eta=\frac{y_1}{\lambda z}$,

$$E(x_1,y_1)=\exp(ikz)\sum a_n\exp\left[\frac{i2\pi nx_1}{L}\right]\exp\left[\frac{-i\lambda zn^2}{L^2}\right]. \tag{69}$$

Note that except the phase term thatis constant over the plane (x_1,y_1), the equation (69) is equal to equation (64) when

$$\exp\left[\frac{-i\lambda zn^2}{L^2}\right]=1. \tag{70}$$

The condition in equation (70) is satisfied when

$$\frac{\lambda z}{L^2}=2m\text{, } m \text{ is an integer.} \tag{71}$$

Consequently, the Talbot panes or the planes where the object is reproduces are located at

$$z = \frac{2mL^2}{\lambda}. \tag{72}$$

7. The LASIK Procedure and Its Problems

Once the aberrations of the eye are determined, they can be corrected using spherocylindrical lenses or by surgery of the eye using LASIK (laser-assisted in situ keratomileusis).LASIK is performed in three steps. The first step is to create a flap of corneal tissue. The second step is to remodel the cornea underneath the flap with the laser. Finally, the flap is repositioned. The problems introduced by LASIK surgery can be classified into three categories:

1. under correction: the procedure doesn't remove enough corneal tissue,
2. overcorrection: the procedure removes too much corneal tissue,
3. corneal wrinkling: the corneal flap created by the microkeratome introduces a small wrinkle or corneal fold when it is replaced, which results in blurred vision in this area [8,9,11,16,19,24,27,36].

An argon fluoride excimer laser operating at wavelength of 193 nm emits light vertically toward the corneal center. The cornea, however, changes in both shape and radius toward the periphery. Light rays that strike an optical surface at a nonperpendicular angle are partially reflected and have less impact than rays that strike at a perpendicular angle. The degree of reflection increases with the degree of inclination of the surface, and energy density decreases as the light rays are spread over a larger surface area. This increasing reflection effect toward the corneal periphery has an impact on the ablation profile of the laser; if this is not taken into account in the design of the ablation profile, a larger amount of tissue will be removed in the center than in the periphery. Under these conditions, a spherical ablation profile will flatten the central cornea to a greater extent than the periphery, creating an oblate ellipsoid with increasing refractive power toward the corneal edges.

In order to eliminate this inconvenience, there is wavefront-guided LASIK, where an ophthalmologist applies a spatially varying correction, guiding the computer-controlled excimer laser using the measurements from a wavefront sensor. The goal is to achieve a more optically perfect eye, though the final result still depends on the physician's success at predicting changes thatoccur during healing [11,19,27,33].

In general, if the wavefront error is given by $W(x,y)$ (i.e., the difference in the shapes of the ideal and aberrated wavefronts), then the thickness $t(x,y)$ of the material that needs to be removed is given by:

$$t(x,y) = (n-1)W(x,y). \tag{73}$$

For custom refractive surgery procedures, the index of refraction, n, will be the index of the cornea. For custom contact lens applications, the index n will be the index of refraction of the contact lens material.

Conclusion

In this chapter, we present some aspects related to visual optics. We focus our attention on colors vision, human eyes' wavefront diseases and correction. Using a spectral image processing algorithm, we simulate the colors perception of a normal viewer. Using the matrix approach, we simulate the vision of the deuteranope and the protanope. In order to characterize the human eyes' wavefront diseases, we present the Zenike polynomials, which represent the mathematical background necessary to study the wavefront aberrations. Also, we can use the formula of Siedel aberrations to compute the monochromatic aberrations. We present some aspects related to Shack-Hartmann sensor, interferograms and Talbot imaging used to capture wavefront aberrations.

Related to these techniques, we present the LASIK surgical procedure by which we can correct the human eye aberrations.In order to correct the low-order wavefront aberrations, people can use eyeglasses made by spherocylindrical lens combinations. However, if they are not satisfied, they can use LASIK or guided LASIK for corrections of the effects of the aberrations.

References

[1] H. Brettel, F. Vienot, J.D. Mollon, *J. Opt. Soc. Am. A*. 14, 10, (1997).

[2] K. Castleman, *Digital image processing*, Prentice Hall, New Jersey (1996).

[3] J. W. Goodman, *Introduction to Fourier optics*, McGraw-Hill, New York (1996).

[4] R. Elkington, J. H. Frank, M. J. Greaney, *Clinical Optics*, Blackwell Science, Bodmin (1999).

[5] R. Fletcher, J. Voke, A. Hilger, *Defective Colour Vision Fundamentals*, *Diagnosis and Management*, Adam Hilger, Bristol, (1995).

[6] R. Gonzales, R.Wood, S. Eddins, *Digital Image Processing Using MATLAB*, Gatesmark Publishing, Upper Saddle River (2009).

[7] B. Mac Evoy, 2010 at http://www.handprint.com/HP/WCL/color1.html.

[8] J. Tyson, 2001 at http://health.howstuffworks.com/lasik.htm.

[9] G.-ming Dai, *Wavefront Optics for Vision Correction*, Spie Press, Bellingham (2008).

[10] S. A. Klein, *J. Opt. Soc. Am. A*, 15, 9 (1998).

[11] R. R. Krueger, R. A. Applegate, S. M. MacRae, *Wavefront Customized Visual Correction The Quest for Super Vision II*, SLACK Incorporated, Thorofare (2004).

[12] H. C. Lee, *Introduction to color imaging science*, Cambridge University Press, Cambridge (2005).

[13] J. Loomis, 2011 at http://www.engr.udayton.edu/faculty/jloomis/courses. html.

[14] D. Malacara, *Optical Shop Testing*, Wiley-Interscience, New Jersey (2007).

[15] D. Malacara, *Color Vision and Colorimetry Theory and Application*, SPIE, Bellingham (2002).

[16] J. M. Miller, R. Anwaruddin, J. Straub, J. Schwiegerling, *J. Refract. Surg.*18, 5 (2002).

[17] N. Ohta, A. R. Robertson, *Colorimetryfundamentals and application*, John Wiley and Sons, New Jersey (2005).

[18] D. Pascale, *A Review of RGB Color Space*, Babel Color Company, Montreal (2003).

[19] J. Porter, H. M. Queener, J. E. Lin, K. Thorn, A. Awwal, *Adaptive Optics for Vision Science Principles, Practices, Design, and Applications*, John Wiley and Sons, New Jersey (2006).

[20] G. Sharma, Ed., *Digital Color Imaging Handbook*, CRC Press, Boca Raton (2003).

[21] J. Schanda Ed., *Colorimetry: Understanding the CIE System*, John Wiley and Sons, New Jersey (2007).

[22] D. Schreuder, *Outdoor Lighting: Physics, Vision and Perception*, Springer, Berlin (2008).

[23] S. K. Shevell, *The science of color*, Elseiver, Oxford (2003).

[24] J. Schwiegerling, *Field Guide to Visual and Opthalmic Optics*, Spie Press, Bellingham, (2004).

[25] L. N. Thibos, X. Hong, *Vis. Sci.* 76, 12 (1999).

[26] F. Toadere, "A method to enhance colors in a digital image," 2008 IEEE-TTTC International Conference on Automation, Quality and Testing, RoboticsAQTR 2008, 22-25, May 2008, (Cluj-Napoca, Romania), pp 262-266.

[27] F. Toadere, "A study about human eyes wavefront aberrations capture and correction," Advancements of Medicine and Health Care through Technology, Meditech 2009, 23-26, Sept. 2009, (Cluj Napoca, Romania), pp 101-104.

[28] F. Toadere, "Colors Visions for a normal Viewer and for a Daltonist,"Advancements of Medicine and Health Care through Technology, Meditech 2009, 23-26 Sept. 2009, (Cluj Napoca), pp 243-246.

[29] F. Toadere, *J. of Phy.: Conf. Series.* 182, 1 (2009).

[30] F. Toadere, "Visual optics an engineering approach," ATOM-N 2010 *The 5th of the international conference advanced topics in optoelectronics, microelectronics and nanotechnologies*, 26-29, Aug. 2010, (Constanța, Romania), pp .

[31] F. Toadere, "A study about daylight perception by the human eyes," *The 3rd WSEAS International Conference on Visualization, Imaging and Simulation,* 3-5, Nov. 2010, (Faro, Portugal), pp 129-134.

[32] H. J. Trussell, M. J. Vrhel, *Fundamental of Digital Imaging*, Cambridge University Press, Cambridge (2008).

[33] Valberg, *Light Vision Color*, John Wiley and Sons Ltd,Chippenham (2005).

[34] P. Viénot, H. Brettel, L. Ott, A. Ben M'Barek, J. D. Mollon, *Nature.* 376, 6536 (1995).

[35] B. Dougherty,A. Wade, 2011 at http://www.vischeck.com/daltonize/.

[36] J. Schwiegerling, 2011 at http://www.visualopticslab.com/opti535.html.

[37] B.Wandell, *Foundation of vision*, Sinauer, Sunderland MA (1995).

[38] W. Welford, *Aberrations of optical systems*, Adam Hilger, Bristol (1986).

[39] J. C. Wyant, 2011 at http://www.optics.arizona.edu/jcwyant/.

In: Medical Physics
Editor: Marcin Balcerzyk

ISBN: 978-1-62257-590-9

Chapter IV

Applications of the Adaptive Wavelet Transform for Analyzing Peripheral Blood Flow Oscillations in the Human Skin

Arina V. Tankanag*

Institute of Cell Biophysics, Russian Academy Science,
Pushchino, Moscow Region, Russia

Abstract

A method for the analysis of peripheral blood flow oscillations is developed, making use of laser Doppler flowmetry (LDF) data and based on continuous wavelet transform and the theory of adaptive wavelets. The capabilities of the developed method have been demonstrated by the examples of sinusoidal model, as well as real LDF signals. The paper shows the method's main advantage proposed over the traditional wavelet analysis, namely, a substantial reduction of "border effects". When a wavelet transform is applied to signals of finite length, border effects can substantially influence the analysis results associated with low-frequency oscillations of peripheral blood flow. A correct wavelet analysis of the latter would, therefore, require a long-term registration of peripheral microhemodynamics. In various studies, the time of LDF signal registration varies from 16 to 30 minutes, causing some difficulties in the study of the microvascular bed. The application of the adaptive wavelets theory should, in our mind, reduce the time of LDF signal recording necessary to assess skin microhemodynamics. A method which reduces information losses at the borders of the signal (particularly, in the initial segment) would be especially important when we are to analyze the low-frequency components of short-living (fast) transitional processes under functional tests. The method of investigating the peripheral blood flow oscillations that we propose – based on adaptive wavelet theory and applying continuous wavelet transform to LDF data – allows one to analyze LDF

* E-mail: tav@icb. psn. ru.

signals recorded during a relatively short time (6 minutes) and do it over a wide frequency range (from 0. 0095 to 2 Hz). Using such a short period of time gives one an advantage in the study of the microvascular bed, as it allows for the reduction of artifacts caused by a prolonged immobilization of the subject during the recording of LDF signals. The technique we propose can be successfully used for the analysis of stationary and nonstationary biomedical signals (ECG, EEG, myogram, etc.), when there is a necessity to investigate time-limited nonstationary rhythmic processes in a wide range of frequencies.

Introduction

The microvascular bed is the smallest functional unit of the circulatory system, in which the transcapillary exchange occurs to support vital tissue homeostasis [11]. In recent years, the method of Laser Doppler Flowmetry (LDF) has become widespread in the studies of peripheral microhemodynamics. The method is associated with an analysis of the Doppler's frequency of a signal shift, obtained upon reflecting the laser beam from erythrocytes [27]. The recorded characteristic of blood flow in the microvascular bed is the microcirculation index, which reflects changes in the perfusion of blood in the probed volume of tissue per time unit and is measured in arbitrary units (a. u.) When the microcirculatory system is investigated using the LDF method, various functional tests are used to measure the adaptive reserves and assess the state of the regulatory mechanisms of skin blood flow [3, 32, 33, 35-37, 40, 46]. So far, all works on microhemodynamics studied with the LDF method have either considered the integral changes in the perfusion of blood in tissues at rest and under functional tests [9, 15, 32, 46, 47], or analyzed the amplitude-frequency spectra of LDF signals [3, 17, 34-37, 40, 48, 49].

It is known that skin blood flow is unstable: it shows temporal and spatial variations. The rhythmic fluctuations of blood flow reflect its variability and adaptability to the ever-changing hemodynamics [38]. They carry information about the state of microhemodynamics, complementary to the commonly used averaged statistical parameters of the microcirculation index. In addition, when the microvascular bed is tested for its reaction to external factors, of great interest is the dynamics of the frequency components of LDF signals, which may reflect the time course of changes in various systems that the peripheral blood flow oscillations depend upon. A method that allows one to isolate specific oscillatory components from the signal analyzed is wavelet filtering. In recent years, the latter has been widely used in the study of various biomedical signals [2, 12, 21, 28, 29, 42, 52].

Numerous LDF-based studies of the microvascular bed have shown that the dynamics of skin blood flow is characterized by several periodic components with characteristic frequencies. The microcirculatory system contains oscillating components, ranging over a wide frequency range, from 0. 0095 to 2 Hz [7, 18, 40]. An application of the traditional wavelet transform is known to have a boundary points problem, which leads to the loss of information at the borders of the signal and can, therefore, affect the analysis results associated with low-frequency oscillations of peripheral blood flow [7, 40]. Consequently, a proper wavelet analysis of low-frequency oscillations of peripheral blood flow would require a long-term registration of peripheral microhemodynamics. In various studies, the time of LDF signal registration varies from 16 to 30 minutes [3, 16, 18, 20, 22, 23, 40], which causes some difficulties in the study of the microvascular bed. We propose a method of amplitude-frequency and time-amplitude analysis of LDF signals, which is based on continuous wavelet transform and adaptive wavelet theory, and allows one to study short-time (6-minute) signals in a wide range of frequencies, from 0. 0095 to 2 Hz [43-45]. The developed method reduces the impact of border effects and, thus, allows one to correctly analyze LDF signals and their frequency components – even when the time of registration is short.

Basic Principles of Wavelet Transform

The wavelet transform of a signal is its expansion in basis soliton-like functions with certain properties. This function is called a wavelet [10]. The general principle for constructing the basis of wavelet transform is to use two operations: dilation and translation.

Using these operations and taking into account the properties of the wavelet locality in the frequency-time domain, one can analyze data at different scales and determine the position of their characteristic features in the time plane. It is due to this scaling that a wavelet analysis is able to reveal differences in the characteristics at different scales, and a translation of wavelet function allows one to analyze the properties of the signal at different points over the entire time segment.

Wavelet transform is used in those cases when the result of signal analysis should contain not only a simple listing of its characteristic frequencies (scales), but also information about specific local coordinates at which these frequencies manifest themselves.

To be generated from the "mother" function by means of translation and dilation is a family of wavelet functions $\psi(t)$:

$$\psi_{a,b}(t)=a^{-1/2}\psi\left(\frac{t-b}{a}\right), \tag{1}$$

where a and b correspond to dilation and translation, respectively.

For a function $\psi(t)$ to be considered a wavelet, it must satisfy certain conditions. First, the function must have a zero mean:

$$\int_{-\infty}^{+\infty}\psi(t)dt=0. \tag{2}$$

Second, the function should be localized in both the physical and Fourier spaces (time and frequency), i. e. , its domains in the physical (Δt) and frequency ($\Delta\omega$) spaces must satisfy the condition $\Delta t\Delta\omega$ = Const $\geq 2\pi$.

The wavelet transform $w(a,t)$ of a signal $f(t)$ is defined as

$$w(a,t)=C_{\psi}^{-1/2}a^{-1/2}\int_{-\infty}^{+\infty}\psi^*\left(\frac{t'-t}{a}\right)f(t')dt', \tag{3}$$

where $\psi(t)$ is a real or complex analyzing wavelet, and * denotes a complex conjugation.

Theory of Adaptive Wavelets

When wavelet analysis is applied to the signals of finite length, there are known to be time delays. These delays are caused by the fact that the center of wavelet, *t,* is at a certain distance from the beginning and end of the signal, which is equal to half of the analyzing wavelet length. Thus, the problem of boundary points may affect the results of analysis on large scales [7, 40].

To calculate wavelet coefficients for the boundary points, we applied the theory of adaptive wavelets, which was designed for analyzing time series with gaps [13, 14].

We believe that an application of adaptive wavelet transform will reduce the influence of boundary points and, thus, improve the efficiency of the wavelet analysis of LDF signals in the low frequency ranges [43-45]. We consider the border of the LDF-signal as a semi-infinite gap. In this case, when the wavelet reaches the border of the signal, it stops to satisfy the admissibility condition (2).

The basic idea of adaptive wavelets is to replace the "corrupted" wavelet with an "adaptive" wavelet $\tilde{\psi}$. The adaptive wavelet must satisfy the admissibility condition $\langle\tilde{\psi}\rangle = 0$ and, with the "gaps" disappearing, must tend to the source wavelet ψ.

According to [13, 14], let us represent the source wavelet ψ as

$$\psi(t) = h(t)\Phi(t), \tag{4}$$

where $\Phi(t)$ is a positively defined scaling function (the Gauss function $\Phi(t) = \exp(-t^2/2)$ is the most convenient) and *h(t)* is the function of wavelet filling. Let us write the unknown function $\tilde{\psi}$ in the form

$$\tilde{\psi}(t',t,a) = [h\left(\frac{t'-t}{a}\right) - C(a,t)]\Phi\left(\frac{t'-t}{a}\right)G(t'), \tag{5}$$

where *G (t)* is the function of "gaps", which is equal to 1 if there is a signal recorded and 0 at all other points (inside discontinuities and outside the signal).

The parameter *C,* depending on the scale factor *a* and the position of the center of the wavelet *t,* is chosen on the basis of the condition $\langle\tilde{\psi}\rangle = 0$ as follows:

$$C(a,t) = \frac{\int_{-\infty}^{+\infty} h(\frac{t'-t}{a})\Phi(\frac{t'-t}{a})G(t')dt'}{\int_{-\infty}^{+\infty} \Phi(\frac{t'-t}{a})G(t')dt'}. \tag{6}$$

Wavelet Filtering

For a wavelet transform (3), an equation exists for an inverse wavelet transform:

$$f(t) = C_{\psi}^{-1} \int_{-\infty}^{+\infty} \int_{-\infty}^{+\infty} a^{-1/2} \psi\left(\frac{t-t'}{a}\right) w(a,t') \frac{dt'da}{a^2} , \qquad (7)$$

where $C_{\psi} = \int_{-\infty}^{+\infty} |\hat{\psi}|^2 |\omega|^{-1} d\omega < \infty$ and $\hat{\psi}(\omega)$ is a Fourier transform of the function $\psi(t)$:

$$\hat{\psi}(\omega) = \int_{-\infty}^{+\infty} \psi(t) e^{-i\omega t} dt . \qquad (8)$$

Thus, the wavelet transform can be regarded as a band-pass filter with a variable bandwidth.

In the procedure of band-pass filtering of finite-length signals (equation (7)), integration limits are replaced with finite values, and, thus, the signal is restored in a given frequency interval:

$$f(t, a_1, a_2) = C_{\psi}^{-1} \int_{a_1}^{a_2} \int_{t_1}^{t_2} a^{-1/2} \psi\left(\frac{t-t'}{a}\right) w(a,t') \frac{dt'da}{a^2} . \qquad (9)$$

It should be noted that the scaling factor a is inversely proportional to frequency.

The Procedure for Analyzing Adaptive Wavelets of Finite-Length Signals

Let us consider the procedure of a non-adaptive wavelet analysis of a signal of the finite length N. The analyzing wavelet is placed at the position W1 (Figure 1), and wavelet coefficients are calculated for the point T1. Then, the analyzing wavelet is shifted along the time axis to the position W2, as shown in Figure 1, and wavelet coefficients are calculated for the point T2.

Thus, when a non-adaptive wavelet transform is used, wavelet coefficients for the initial segment of the signal (i. e., for the time points $0 < t <$ T1) are either not defined or equal to zero.

An analogous situation takes place at the right border of the signal: Tn - 1 and Tn are the next to last and last points for which wavelet coefficients will be calculated. For the points Tn $< t <$ N, wavelet coefficients are either not defined or equal to zero (Figure 1).

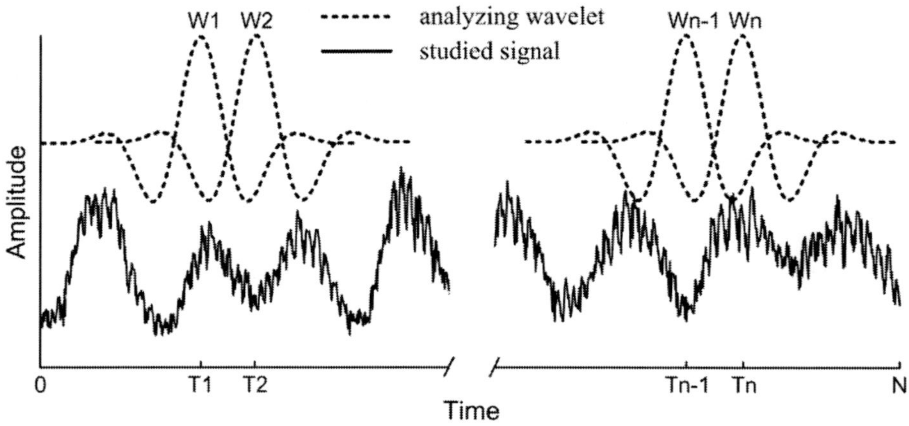

Figure 1. A scheme for non-adaptive wavelet analysis.

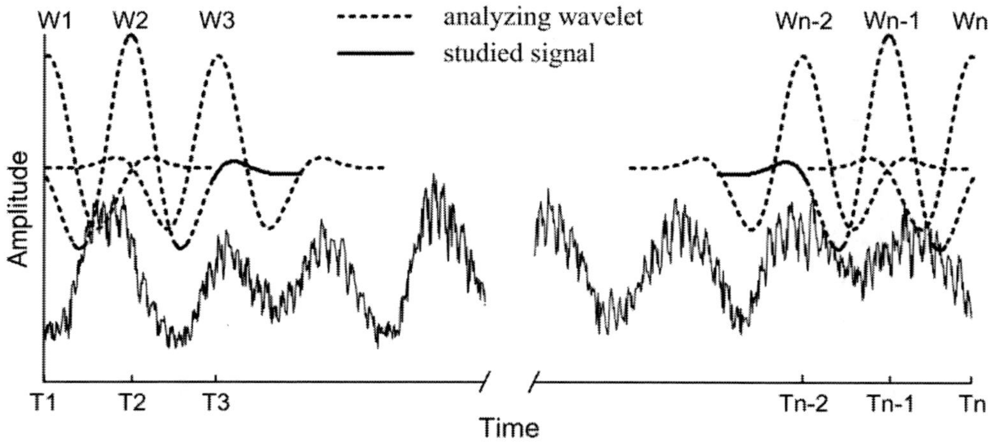

Figure 2. A scheme for adaptive wavelet analysis.

The loss of information at the signal borders gives rise to boundary effects, which, in turn, will affect the analysis results associated with the low-frequency oscillations of the signal [7, 40]. A correct wavelet analysis of low-frequency oscillations of peripheral blood flow will, therefore, require the registration of peripheral microhemodynamics to be rather long. The recommended time of recording must be long enough for the signal to cover at least 10 periods of the lowest frequency oscillatory component [6]. In various studies, the time of LDF signal registration varies from 16 to 30 minutes [3, 16, 18, 20, 22, 23, 40], which creates some difficulties in examining the microvascular bed.

The proposed modification of wavelet transform, which is based on the theory of adaptive wavelets described above [13, 14], allows one to minimize border effects, eliminating the influence of boundary points on large scales. An adaptive wavelet analysis is performed according to the following algorithm: The analyzing wavelet is shifted so that its center is set at the starting point of the signal (position W1 in Figure 2). Then, the coefficient $C(a, b)$ is calculated according to (6), followed by the construction of a modified (adaptive) wavelet $\widetilde{\psi}$ according to (5) and calculation of the coefficients $W(a, b, t')$ for the time point T1. Such an adaptation occurs every time the analyzing wavelet is shifted (position W2 in

Figure 2) – until the starting point of the analyzing wavelet gets to the beginning of the signal (position W3 in Figure 2). After this, a non-adaptive wavelet analysis is applied – until the analyzing wavelet is brought to the position Wn - 2 in Figure 2. At the end of the signal, the wavelet is adapted again (the positions Wn - 1 and Wn in Figure 2). Thus, wavelet coefficients will be calculated for all points of the analyzed signal (T1 < t < Tn). The adaptation is applied for each scale a. The detailed algorithm for calculating wavelet coefficients is given in the Appendix. The proposed modification of wavelet transform allows one to use all the information contained in the signal. Thus, using the adaptive wavelet analysis makes it possible to reduce the border effects and allows one to perform a more accurate analysis of low-frequency oscillations in short LDF signals. It should be noted that in the spectral analysis, the adaptation of mother wavelets is performed once, whereas wavelet filtering does it twice: the first time, in the procedure of the wavelet coefficients calculation according to (3); and a second time, in the process of signal reconstruction in a given frequency interval according to (9).

Table 1. Frequency intervals

Interval	Frequency (Hz)
I	0. 0095-0. 021
II	0. 021-0. 056
III	0. 056-0. 145
IV	0. 145-0. 6
V	0. 6-2

An Adaptive Wavelet Analysis of Model Sinusoidal Signals

To demonstrate the capabilities of the method proposed, we analyzed a model signal with known amplitude and frequency characteristics. The model signal was chosen as the sum of 5 sines with the frequencies 0. 01, 0. 03, 0. 1, 0. 3 and 1 Hz and amplitudes equal to 1. 5 a. u. The frequencies were chosen so to correspond to the center frequencies of the oscillatory components at the microcirculatory level, which reflect the regulatory processes of the cardiovascular system [40]. In the frequency band from 0. 0095 to 2 Hz, five non-overlapping intervals are identified (Table 1). In these intervals, oscillations of blood flow are determined by the influence of endothelial [3, 18, 40, 41], neurogenic [19], myogenic [4], respiratory [5, 26] and cardiac [7, 40] processes, respectively. It is now actively discussed whether the microvascular bed contains a low-frequency (centered at ~ 0. 007 Hz) NO-independent component [1, 3, 18]. The present issue will not consider this range.

As the analyzing wavelet function, we took the real part of the complex Morlet wavelet, which is widely used in the study of LDF signals [3, 16, 20, 22, 23, 40]:

$$\psi(t) = e^{i\omega_0 t} e^{(-t^2/2)} \quad (10)$$

The choice ω_0 represents a compromise between the localization of the wavelet in either the time or frequency domain. Smaller values of ω_0 are used to study low-frequency components of the initial signal; large values, to study high-frequency components [7].

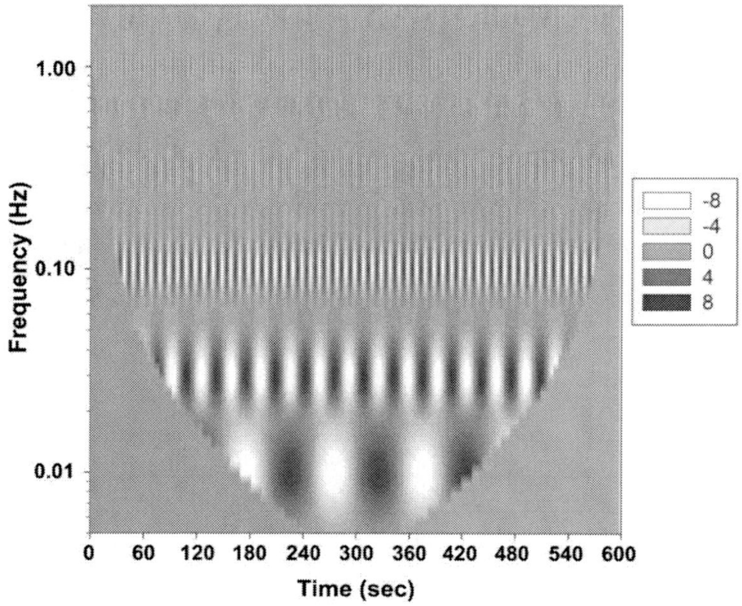

Figure 3. A time-frequency representation of non-adaptive wavelet coefficients of the model signal. The legend shows values of wavelet coefficients.

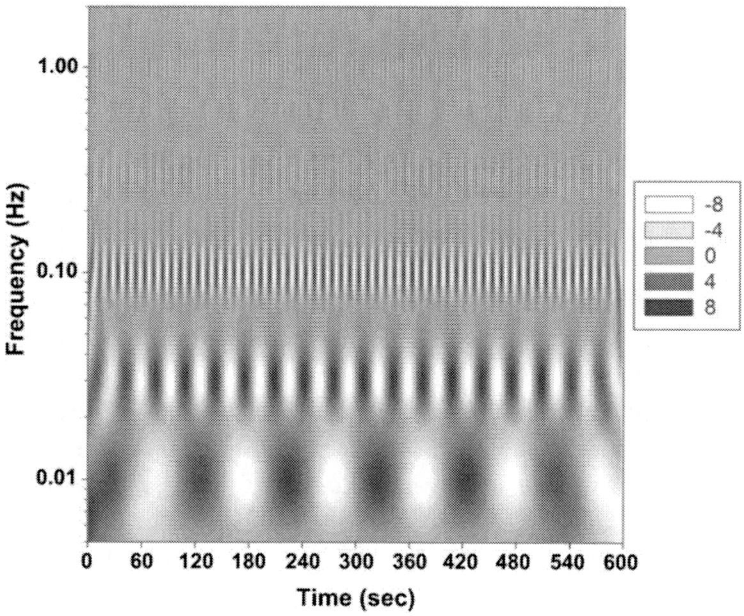

Figure 4. A time-frequency representation of adaptive wavelet coefficients of the model signal. The legend shows values of wavelet coefficients.

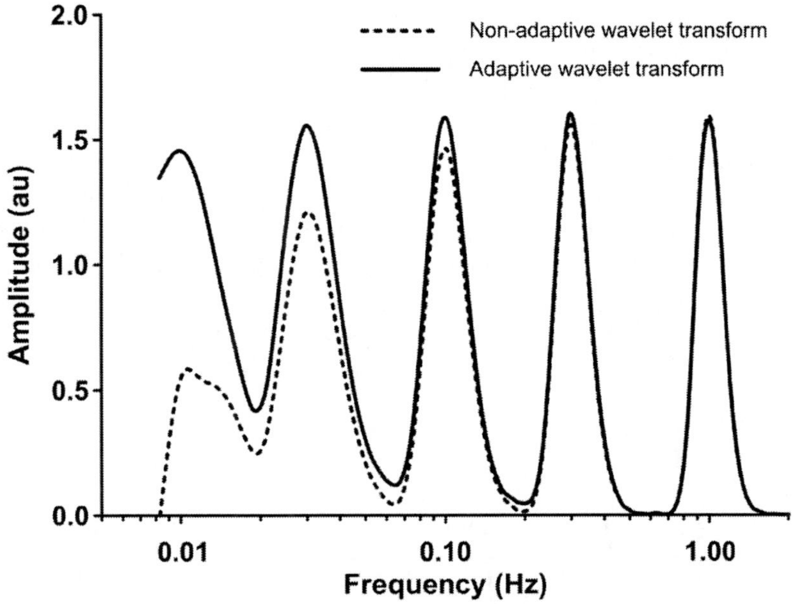

Figure 5. Amplitude-frequency spectra of the model 6-minute signal.

As a result of the non-adaptive wavelet analysis of the model signal, a three-dimensional array of wavelet coefficients, *W(a, t)*, was obtained. Figures 3 and 4 show projections of arrays of non-adaptive and adaptive wavelet coefficients on the "time-frequency" plane, respectively. Frequency is inversely proportional to the scale *a*. For better presentation, the frequency (scale) axis in the figures is logarithmic, and the time scale is linear. Such a presentation of the wavelet analysis results gives a visual qualitative picture (map) of the temporal distribution of oscillation amplitudes over the entire frequency range.

Figure 3 shows that the use of non-adaptive wavelet transform resulted in a significant decrease of the values of wavelet coefficients at low frequencies. On the contrary, using the adaptive wavelet analysis enabled us to calculate wavelet coefficients over the entire signal and, thereby, to significantly reduce the loss of information at the borders of the signal (Figure 4).

Nowadays, the researchers studying the oscillations of blood flow velocity in the microvascular bed widely use a spectral analysis, which allows one to evaluate the amplitude-frequency spectra of signals [16, 17, 31-37, 40, 43, 48, 49]. Correspondingly, we conducted non-adaptive and adaptive spectral analyses of a model 6-min signal (Figure 5).

When a non-adaptive wavelet transform was used, the amplitude-frequency spectrum showed a significant, compared to the results of adaptive wavelet analysis, decrease of oscillation amplitudes at low frequencies (below 0. 1 Hz; Figure 5). For comparison, Figure 6 shows non-adaptive and adaptive amplitude-frequency spectra of a 20-minute model signal. The time of 20 minutes was chosen according to the recommendations by [6], which state that the recording must be as long as to contain at least 10 periods of the lowest frequency component of the signal. For a signal of this length, there are almost no differences between non-adaptive and adaptive wavelet spectra (Figure 6). It should be noted that the adaptive spectrum of the 6-minute model signal (Figure 5) is almost identical to the non-adaptive and adaptive spectra of the 20-minute model signal (Figure 6). We thereby suppose that the use of adaptive wavelets enables one to conduct a more correct spectral analysis of short signals (~6 minutes long). Such a short recording time allows one to analyze the signals registered in patients who had difficulties lying motionless during the experiment and, hence, to avoid artifacts at the recording stage.

It is known that, when studying the response of the microvascular bed to external factors, researchers pay much attention to the dynamics of the frequency components of LDF signals, which can reflect the time course of changes in the systems that determine the oscillations of peripheral blood flow. One of the methods that are able to isolate specific oscillatory components from a signal is band-pass wavelet filtering, which has been widely used in recent years in the study of various biomedical signals [2, 12, 21, 28, 29, 42, 52].

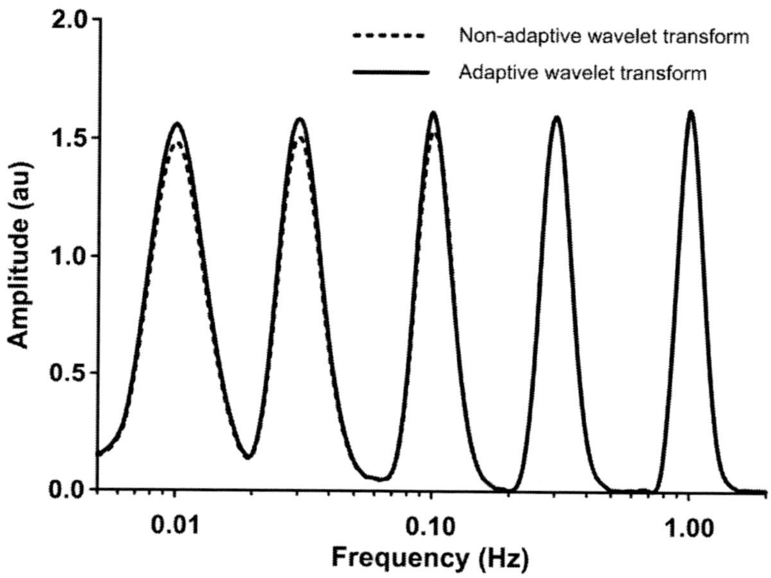

Figure 6. Amplitude-frequency spectra of the model 20-minute signal.

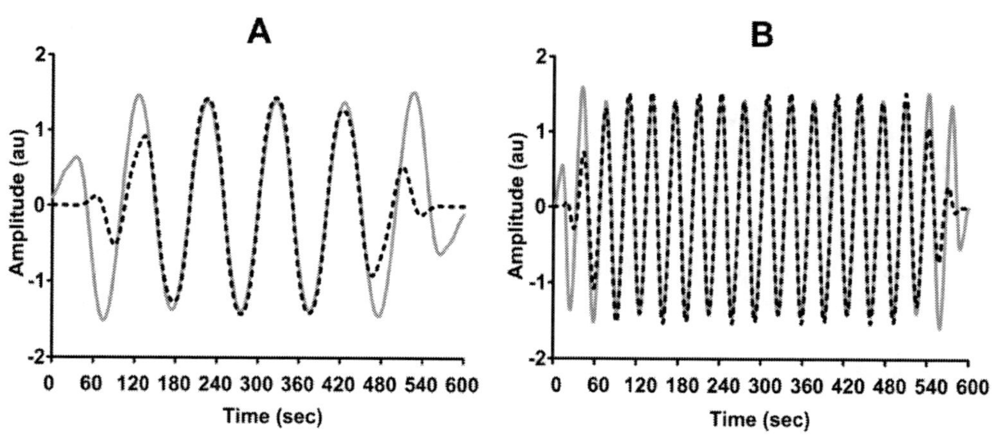

Figure 7. Oscillatory components of the model 10-minute signal filtered in the I (A) and II (B) intervals: non-adaptive (black dashed line) and adaptive (gray solid line) wavelet filtering.

We believe that the use of adaptive wavelets would improve the efficiency of wavelet filtering. To confirm this hypothesis, we compared the results of non-adaptive and adaptive band-pass wavelet filtering of a model 10-minute signal (Figure 7). As seen from Figure 7, non-adaptive wavelet filtering greatly reduces the amplitudes of the recovered signal at the borders. It should also be noted that the border effects are more pronounced when the source signal is filtered in the low-frequency region.

In the case of the I interval, for example, the characteristic reduction of oscillation amplitudes after non-adaptive wavelet filtering occurs over most parts of the analyzed time segment (Figure 7A).

For the oscillatory components of higher frequencies, the border regions distorted by the boundary effects are shorter (Figure 7B). It should be noted that the influence of border effects is reduced as the frequency of oscillating components grows. At the same time, adaptive wavelet filtering allows one to restore the model signal in the specified frequency interval without significant loss of information at the signal borders (Figure 7). The results we obtained are of particular importance when the low-frequency components of short LDF signals are studied when there are only a few oscillation periods of a given frequency interval in the segment analyzed.

An Adaptive Wavelet Analysis of LDF Signals

The method that reduces information losses at the signal borders (particularly, in the initial segment) is especially useful when we are to analyze low-frequency components of short (fast) transitional processes under functional tests. Such tests, particularly the occlusion test [9, 25, 30, 32-34, 37, 46-48, 51, 53], have become widespread in the studies of peripheral blood flow by the LDF method. The occlusion test is implemented by clamping a limb for 1-3 minutes with the tonometer cuff to stop the blood flow and induce ischemia in the area examined. After occlusion is removed, the blood flow is restored, and a post-occlusion reactive hyperemia is developed, which is manifests itself in the enhancement of skin blood flow to a value greater than that at rest, with the subsequent subsiding of the flow to the initial level [9, 47]. The analysis of blood perfusion dynamics in the examined area during post-occlusion reactive hyperemia provides information on changes in the microvascular bed, and this information can be used as a diagnostic criterion in the study of various vascular diseases [8, 24]. The reaction of blood flow to short-term ischemia is estimated on the basis of a large number of parameters [9, 25, 30, 32-34, 53]. Some works are devoted to the study of integral amplitude-frequency characteristics of peripheral blood flow during the occlusion test [30, 33, 34, 37, 48]. The spectral analysis of peripheral blood flow oscillations provides information on the time-averaged changes of oscillation amplitudes and does not allow one to study the amplitude-temporal characteristics of the oscillatory components of LDF signals. We propose a method that makes it possible to study nonstationary oscillatory processes occurring in the microvascular bed after an occlusion break.

The cutaneous blood perfusion was measured on the outer surface of the right forearm with a laser Doppler flowmeter LAKK-02 ("LAZMA", Russia) for 14 minutes. The sampling rate was 20 Hz. Figure 8 shows the post-occlusion fragment of the LDF signal, corresponding to the interval from the occlusion break to the end of measurements (gray line).

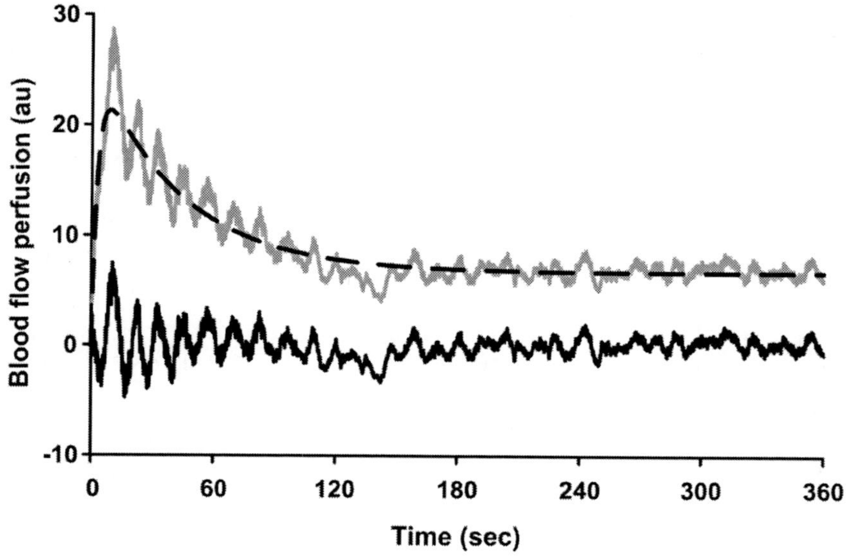

Figure 8. The post-occlusion fragment of the LDF signal (gray line), the two-exponential regression function (black dashed line) and the oscillating signal obtained by subtracting the regression function from the postocclusive LDF-signal fragment (black solid line).

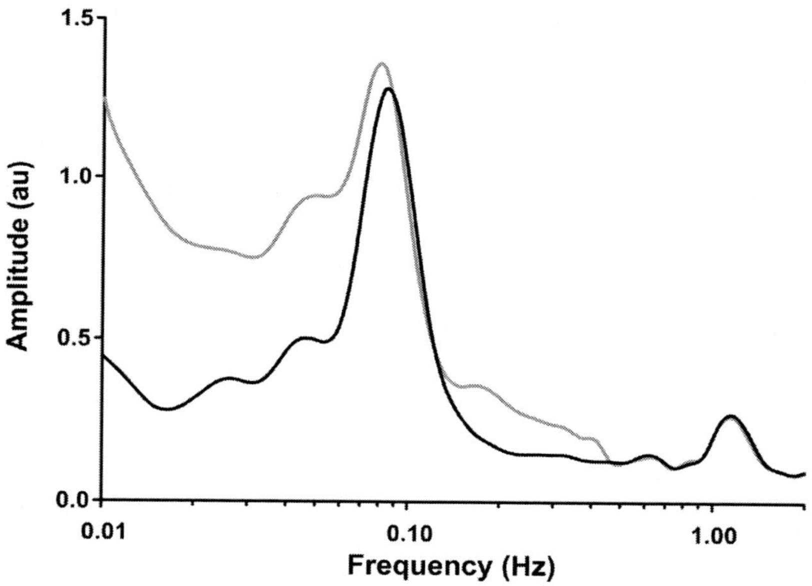

Figure 9. Amplitude-frequency spectra of the initial post-occlusion LDF-signal fragment (gray line) and linearized oscillating signal (black line).

As seen from Figure 8, the microvascular bed reaction to the stop of ischemia is characterized by two phases: (1) a sharp increase in blood flow at the moment of occlusion removal; and (2) the development of post-occlusion reactive hyperemia followed by the restoration of blood flow. Against the background of blood flow rise and decline, one can see marked oscillations.

Figure 9 shows an amplitude-frequency spectrum of the source post-occlusion fragment of the LDF signal. As seen in the figure, the amplitude of blood flow oscillations in the low-frequency region is substantially increased. Thus, the process of sharp rise of blood flow perfusion followed by its decline makes significant distortions in the amplitude-frequency spectrum of the post-occlusion LDF signal.

We suggested that these distortions can be avoided if we formally represent the post-occlusion signal as a sum of two separate processes: (1) the overall trend of perfusion changes and (2) the dynamics of the signal oscillatory components. The first process describes the dynamics of hyperemia; the second one characterizes the oscillatory phenomena that accompany hyperemia.

In the example considered, the dynamics of perfusion can be described by two exponents, the first of which characterizes the rise of perfusion and the second corresponds to its decline to the initial level. Figure 8 shows a double-exponential regression function (black dashed line) that fits the post-occlusion fragment of the LDF signal (gray solid line) with a high degree of correlation (R = 0. 95). Thus, by subtracting this regression function from the post-occlusion LDFsignal fragment, we obtained an oscillating signal that preserves the frequency components corresponding to the oscillations of the source post-occlusion signal (Figure 8, black solid line).

Figure 9 shows an amplitude-frequency spectrum of this linearized oscillating signal. With the regression function subtracted from the post-occlusion signal fragment (linearization), the amplitude of low-frequency oscillatory components becomes, as the figure 9 shows, significantly lower. Thus, the procedure of linearization allows one, in our opinion, to perform a more correct analysis of oscillatory components of LDF signals – those signals that have a strong non-linear trend (i. e. , the signals recorded under functional tests).

In the case of spectral analysis of peripheral blood flow oscillations, it should be noted that the latter only gives information on the time-averaged changes of oscillation amplitudes and does not allow one to study the amplitude-temporal characteristics of the oscillatory components of LDF signals.

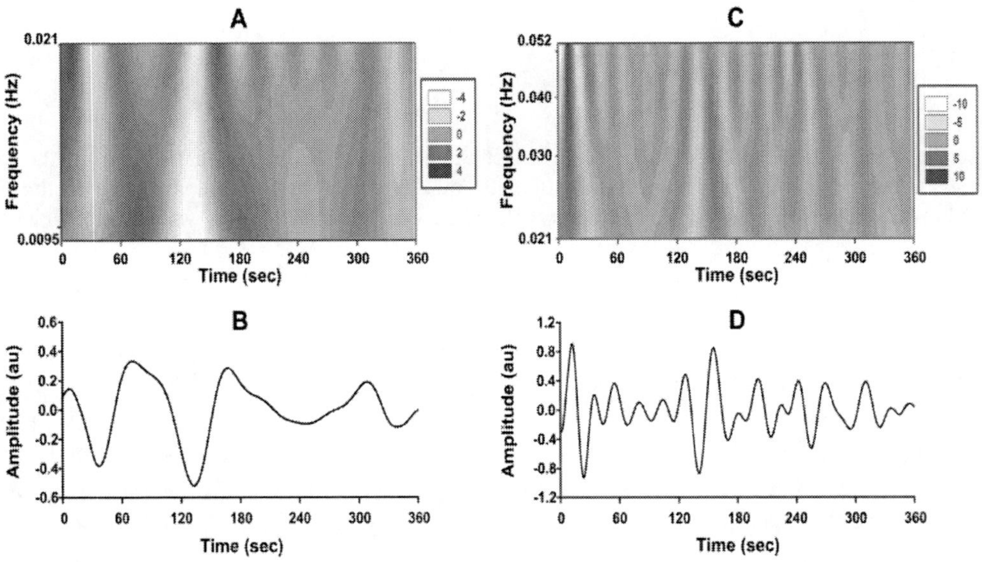

Figure 10. A time-frequency representation of wavelet coefficients of the post-occlusive oscillating LDF signal for the I (A) and II (C) intervals and the oscillatory components of blood flow filtered in these intervals (B and D, respectively). The legends show values of wavelet coefficients.

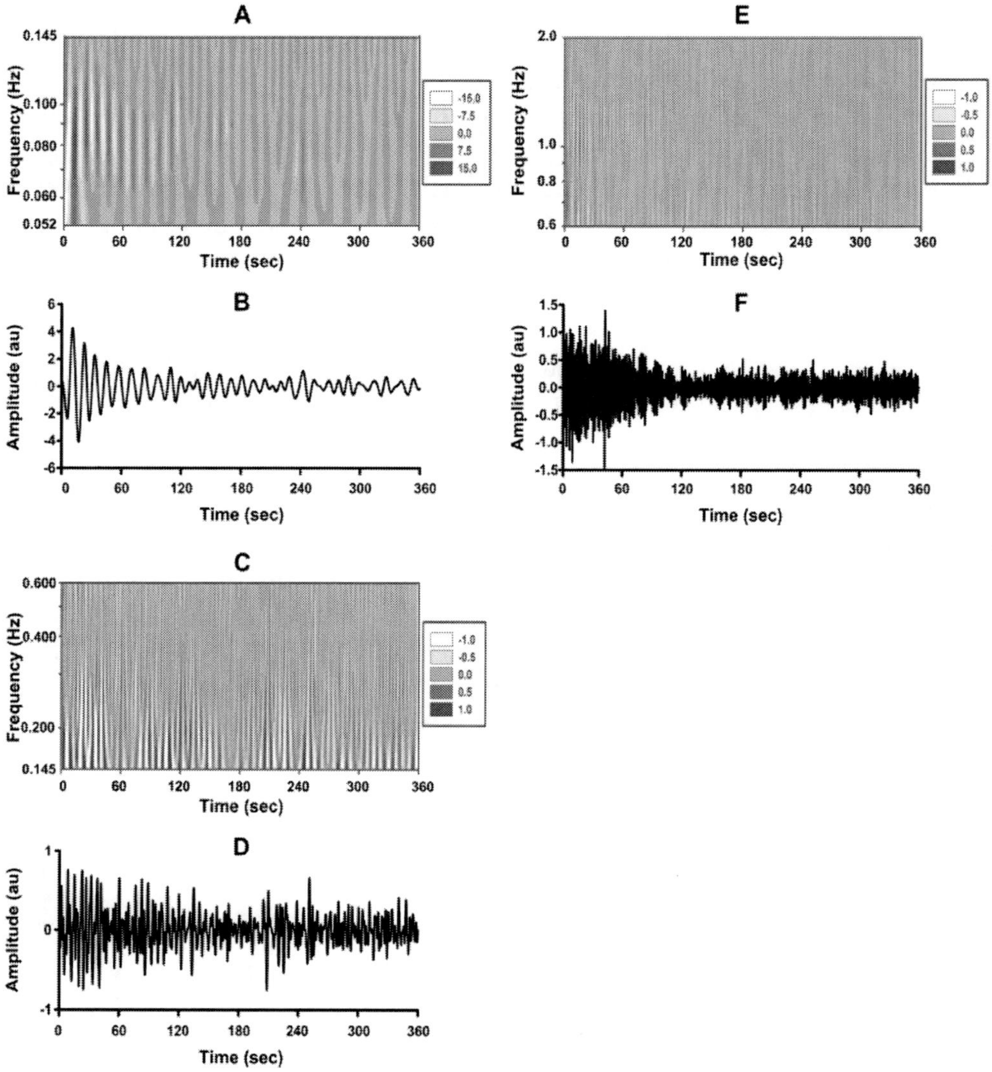

Figure 11. A time-frequency representation of wavelet coefficients of the post-occlusion oscillating LDF signal for the III (A), IV (C) and V (E) intervals and the oscillatory components of blood flow filtered in these intervals (B, D and F, respectively). The legends show values of wavelet coefficients.

To analyze the dynamics of peripheral blood flow oscillations, we subjected the oscillating signal obtained to the procedure of adaptive wavelet filtering in the I, II (Figure 10), III, IV and V (Figure 11) frequency intervals. For clarity, we also presented, for each interval, the corresponding fragments of the projections of the three-dimensional array of wavelet coefficients.

As seen in Figure 10, the wavelet coefficients for the I and II intervals (Figure 10A and 10C, respectively) and the amplitudes of skin blood flow oscillations are characterized by the absence of a visible reaction to forearm ischemia. In the III (Figure 11A) and V (Figure 11E) intervals, though, removal of ischemia was accompanied by a significant increase in the values of wavelet coefficients, followed by their diminishing to the level close to that at rest. In the IV interval (Figure 11C), we registered no such marked an increase in the wavelet coefficients after the cuff release.

Figures 11B, 11D and 11F show the results of adaptive wavelet filtering of blood flow oscillations in the III, IV and V intervals, respectively. As seen from Figure 11, the amplitude of oscillaitons in these intervals rises significantly after ischemia removal and then declines to the initial level. For the III interval, the maximal amplitude amounted to ~4 a. u. ; for the IV and V intervals, it was ~1 a. u. Thus, the proposed method of adaptive wavelet filtering allows one to quantify changes of the amplitude of blood flow oscillations in different frequency intervals.

Conclusion

The developed method of adaptive wavelet transform makes it possible to perform a spectral analysis and to study the dynamics of oscillations of the LDF signal amplitude in different frequency intervals. The application of adaptive wavelets reduces the effect of boundary points, which is especially important when the amplitude of low-frequency oscillatory components of peripheral blood flow is studied. The method proposed can be successfully used both to examine the dynamics of amplitude oscillations in a state of rest and to study the reaction of the microcirculatory system to various functional tests. The method is particularly useful when there is a necessity to analyze the low-frequency components of short (fast) transitional processes under functional tests. Furthermore, the application of adaptive wavelet transform reduces the time of LDF signal registration. This advantage can be useful in clinical practice, as it decreases the time during which the patient should remain motionless and, hence, significantly reduces the number of recording artifacts.

The method proposed was successfully used in the study of age-related features of the microvascular bed functioning [47, 48, 50], and in the studies that examined the reaction of the microcirculatory system in conditionally healthy volunteers to various functional tests [17, 49, 51]. Currently, adaptive wavelets are being applied for the analysis of ultradian and circadian rhythms in the dynamics of skin surface temperature [39]. We believe that the method proposed can be used to filter various stationary and nonstationary biomedical signals (ECG, EEG, myogram, etc.), as well as signals studied in other fields of science and technology.

Acknowledgments

The work is supported by the Russian Foundation for Basic Research (RFBR grant #09-04-00902).

Appendix

Given below is the algorithm for adaptive wavelet analysis. *F*, a source signal with the number of elements *N*. *Wl*, the analyzing wavelet with the number of elements *WlN*. Φ, the scale function; *h*, the filling function. *Wk*, an array of wavelet coefficients with the number of elements *N*. The scheme shows a single iteration for the scale *a*. The algorithm is implemented in Delphi (Borland).

Step 1. $t = 1$.

Step 2. The parameter *C* is calculated

$$C = \frac{\sum_{i=1}^{WIN} h(i-t)\hat{O}(i-t)G(i)}{\sum_{i=1}^{WIN} \hat{O}(i-t)G(i)},$$

where

$$G(i) = \begin{cases} 0, \text{ if } 1 \le i \le \frac{WIN}{2} - t. \\ 1, \text{otherwise} \end{cases}$$

Step 3. The adaptive analyzing wavelet is computed

for i = 1 to WIN: $Wl(i) = [h(i-t) - C]\Phi(i-t)G(i)$, where

$$G(i) = \begin{cases} 0, \text{ if } 1 \le i \le \frac{WIN}{2} - t \\ 1, \text{otherwise} \end{cases}.$$

Step 4. The wavelet coefficient *Wk(t)* is calculated using the adaptive analyzing wavelet *Wl*

$$Wk(t) = \sum_{i=1}^{\frac{WIN}{2}+t} Wl(i)F(i+t-\frac{WIN}{2}).$$

Step 5. $t = t + 1$.

Step 6. If $t \le \frac{WIN}{2} - 1$ then go to step 2, else go to step 7.

Step 7. $t = \frac{WIN}{2}$.

Step 8. $C = \frac{\sum_{i=1}^{WIN} h(i-t)\hat{O}(i-t)}{\sum_{i=1}^{WIN} \hat{O}(i-t)}$.

Step 9. The non-adaptive analyzing wavelet is computed for

$i = 1$ to WlN: $Wl(i) = [h(i-t) - C]\Phi(i-t)$.

Step 10. $Wk(t) = \sum_{i=1}^{WlN} Wl(i)F(i+t-\frac{WlN}{2})$.

Step 11. $t = t + 1$.

Step 12. If $t \le N - \frac{WlN}{2}$ then go to step 8, else go to step 13.

Step 13. $t = N - \frac{WlN}{2} + 1$.

Step 14.

$$C = \frac{\sum_{i=1}^{WlN} h(i-t)\,\hat{O}(i-t)G(i)}{\sum_{i=1}^{WlN} \hat{O}(i-t)G(i)},$$

where

$$G(i) = \begin{cases} 0, \text{ if } N - t + \frac{WlN}{2} \le i \le WlN \\ 1, \text{otherwise} \end{cases}.$$

Step 15. The adaptive analyzing wavelet is computed for
$i = 1$ to WlN: $Wl(i) = [h(i-t) - C]\Phi(i-t)G(i)$, where

$$G(i) = \begin{cases} 0, \text{ if } N - t + \frac{WlN}{2} \le i \le WlN \\ 1, \text{otherwise} \end{cases}.$$

Step 16.

$$Wk(t) = \sum_{i=1}^{\frac{WlN}{2}+t} Wl(i)F(i+t-\frac{WlN}{2}).$$

Step 17. $t = t + 1$.

Step 18. If $t \le N$ then go to step 14, else the iteration is ended.

References

[1] K. Azman-Juvan, A. Bernjak, V. Urbancic-Rovan, A. Stefanovska and D. Stajer, *J. Vasc. Res.* 45, 164 (2008).

[2] O. Bai, P. Lin, S. Vorbach, J. Li, S. Furlani and M. Hallett, *Clin. Neurophysiol.* 118, 2637 (2007).

[3] A. Bernjak, P. B. Clarkson, P. V. McClintock and A. Stefanovska, *Microvasc. Res.* 76, 224 (2008).

[4] S. Bertuglia, A. Colantuoni and M. Intaglietta, *Microvasc. Res.* 48, 68 (1994).

[5] A. Bollinger, A. Yanar, U. Hoffmann and U. K. Franzeck, *Is high-frequency fluxmotion due to respiration or to vasomotion activity?* In: K. Messmer (Ed.), "Progress in Applied Microcirculation", Karger, Basel 52-58 (1993).

[6] M. B. Bracic Lotric, A. Stefanovska, D. Stajer and V. Urbancic-Rovan, *Physiol. Meas.* 21, 441 (2000).

[7] M. Bracic and A. Stefanovska, *Bull. Math. Biol.* 60, 919 (1998).

[8] P. A. Carberry, A. M. Shepherd and J. M. Johnson, *Hypertension* 20, 349 (1992).

[9] J. L. Cracowski, C. T. Minson, M. Salvat-Melis and J. R. Halliwill, *Trends Pharmacol. Sci.* 27, 503 (2006).

[10] I. Daubechies, *Ten Lectures on Wavelets*, SIAM, Philadelphia (1992).

[11] B. Fagrell and M. Intaglietta, *J. Intern. Med.* 241, 349 (1997).

[12] G. Foca, C. Ferrari, N. Sinelli, M. Mariotti, M. Lucisano, R. Caramanico and A. Ulrici, *Anal. Bioanal. Chem.* 399, 1965 (2011).

[13] P. Frick, S. L. Baliunas, D. Galyagin, D. Sokoloff and W. Soon, *Astrophys. J.* 483, 426 (1997).

[14] P. Frick, A. Grossmann and P. Tchamitchian, *J. Math. Phys.* 38, 4091 (1998).

[15] A. Humeau, J. L. Saumet and J. P. L'Huillier, *Microvasc. Res.* 60, 141 (2000).

[16] A. Humeau, A. Koitka, P. Abraham, J. L. Saumet and J. P. L'Huillier, *Phys. Med. Biol.* 49, 843 (2004).

[17] G. V. Krasnikov, A. V. Tankanag, T. N. Koniaeva, T. N. Piskunova and N. K. Chemeris, *Ross. Fiziol. Zh. Im. I. M. Sechenova* 93, 394 (2007) (In Russian).

[18] P. Kvandal, S. A. Landsverk, A. Bernjak, A. Stefanovska, H. D. Kvernmo and K. A. Kirkeboen, *Microvasc. Res.* 72, 120 (2006).

[19] S. A. Landsverk, P. Kvandal, T. Kjelstrup, U. Benko, A. Bernjak, A. Stefanovska, H. Kvernmo and K. A. Kirkeboen, *Anesthesiology* 105, 478 (2006).

[20] S. A. Landsverk, P. Kvandal, A. Bernjak, A. Stefanovska and K. A. Kirkeboen, *Anesth. Analg.* 105, 1012 (2007).

[21] L. Lehtola, M. Karsikas, M. Koskinen, H. Huikuri and T. Seppanen, *J. Med. Eng. Technol.* 32, 400 (2008).

[22] Z. Li, E. W. Tam, M. P. Kwan, A. F. Mak, S. C. Lo and M. C. Leung, *Phys. Med. Biol.* 51, 2681 (2006).

[23] X. Liu, B. Zeng, C. Fan, P. Jiang and X. Hu, *Phys. Med. Biol.* 51, 173 (2006).

[24] T. F. Luscher and G. Noll, *J. Hypertens. Suppl.* 14, S111-S119 (1996).

[25] F. Morales, R. Graaff, A. J. Smit, S. Bertuglia, A. L. Petoukhova, W. Steenbergen, P. Leger and G. Rakhorst, *Microvasc. Res.* 69, 17 (2005).

[26] M. E. Muck-Weymann, H. P. Albrecht, D. Hager, D. Hiller, O. P. Hornstein and R. D. Bauer, *Microvasc. Res.* 52, 69 (1996).
[27] P. A. Oberg, T. Tenland and G. E. Nilsson, *Acta Med. Scand. Suppl.* 687, 17 (1984).
[28] M. T. Pourazad, Z. Moussavi and G. Thomas, *Med. Biol. Eng. Comput.* 44, 216 (2006).
[29] Y. Ren, M. T. Johnson and J. Tao, *J. Acoust. Soc Am.* 124, 316 (2008).
[30] M. Rossi, A. Carpi, M. C. Di, F. Galetta and G. Santoro, *Microvasc. Res.* 72, 34 (2006).
[31] M. Rossi, A. Carpi, M. C. Di, F. Galetta and G. Santoro, *Microvasc. Res.* 73, 124 (2007).
[32] M. Rossi, A. Carpi, M. C. Di, F. Franzoni, F. Galetta and G. Santoro, *Microvasc. Res.* 74, 9 (2007).
[33] M. Rossi, A. Carpi, M. C. Di, F. Galetta and G. Santoro, *Clin. Hemorheol. Microcirc.* 36, 163 (2007).
[34] M. Rossi, A. Cupisti, M. C. Di, F. Galetta, G. Barsotti and G. Santoro, *Microvasc. Res.* 75, 315 (2008).
[35] M. Rossi, A. Carpi, M. C. Di, F. Franzoni, F. Galetta and G. Santoro, *Physiol. Res.* 58, 39 (2009).
[36] M. Rossi, M. C. Di, P. Erba, F. Galetta, A. Carpi and G. Santoro, *Clin. Hemorheol. Microcirc.* 42, 107 (2009).
[37] M. Rossi, M. Nannipieri, M. Anselmino, M. Pesce, E. Muscelli, G. Santoro and E. Ferrannini, *Obes. Surg.* 21, 87 (2011).
[38] J. A. Schmidt, P. Borgstrom, G. P. Firestone, W. P. von, M. Intaglietta and A. Fronek, *J. Vasc. Surg.* 18, 207 (1993).
[39] C. G. Scully, A. Karaboue, W. -M. Liu, J. Meyer, P. F. Innominato, K. H. Chon, A. M. Gorbach and F. Levi, *Interface Focus*1, 48 (2011).
[40] A. Stefanovska, M. Bracic and H. D. Kvernmo, *IEEE Trans. Biomed. Eng.* 46, 1230 (1999).
[41] J. M. Stewart, I. Taneja, M. S. Goligorsky and M. S. Medow, *Microcirculation* 14, 169 (2007).
[42] Y. Su and K. I. Shoghi, *Phys. Med. Biol.* 53, 5899 (2008).
[43] A. Tankanag and N. Chemeris, *Phys. Med. Biol.* 53, 5967 (2008).
[44] A. V. Tankanag and N. K. Chemeris, *Biofizika* 54, 537 (2009).
[45] A. V. Tankanag and N. K. Chemeris, *Phys. Med. Biol.* 54, 5935 (2009).
[46] G. A. Tew, M. Klonizakis and J. M. Saxton, *Eur. J. Appl. Physiol.* 109, 173 (2010).
[47] I. V. Tikhonova, A. V. Tankanag, N. I. Kosiakova and N. K. Chemeris, *Ross. Fiziol. Zh. Im. I. M. Sechenova* 91, 1132 (2005) (In Russian).
[48] I. V. Tikhonova, A. V. Tankanag, N. I. Kosiakova and N. K. Chemeris, *Ross. Fiziol. Zh. Im. I. M. Sechenova* 91, 1305 (2005) (In Russian).
[49] I. V. Tikhonova, A. V. Tankanag, N. I. Kosiakova and N. K. Chemeris, *Ross. Fiziol. Zh. Im. I. M. Sechenova* 92, 1429 (2006) (In Russian).
[50] I. V. Tikhonova, A. V. Tankanag and N. K. Chemeris, *Fiziol. Cheloveka* 36,114 (2010).
[51] I. V. Tikhonova, A. V. Tankanag and N. K. Chemeris, *Microvasc. Res.* 80, 58 (2010).
[52] A. B. Wiltschko, G. J. Gage and J. D. Berke, *J. Neurosci. Methods* 173, 34 (2008).
[53] G. B. Yvonne-Tee, A. H. Rasool, A. S. Halim and A. R. Rahman, *J. Pharmacol. Toxicol. Methods* 52, 286 (2005).

In: Medical Physics
Editor: Marcin Balcerzyk
ISBN: 978-1-62257-590-9

Chapter V

Applications of the Monte Carlo Method in Medical Physics

Ernesto Amato[1], Domenico Lizio[2] and Sergio Baldari[1]
[1]Department of Radiological Sciences, University of Messina, Italy
[2]Research Center of Saluggia, Institute of Radiological Protection, ENEA, Italy

Abstract

The Monte Carlo simulation of particle transport and interaction in matter finds growing applications in medical radiation physics. Dosimetric applications in radiation therapy span from internal dosimetry in radionuclide therapy of nuclear medicine, to the treatment planning in external beam radiation therapy with photons, electrons or fast heavy ions, to the assessment of radiation dose distribution in heterogeneous media such as lungs, bones or renal parenchyma.

After a general introduction on the different Monte Carlo packages available (Geant4, Fluka, EGS, MCNP, etc.), we focus on the validation of physical models and parametrizations mostly relevant in the energy range of interest for medical applications, and on the comparison of the outcomes from the various simulation packages.

We then report on the recent applications of such packages in the development of new and more accurate models for external beam radiation therapy with photons and electrons, hadron therapy applications, and the internal dosimetry of radionuclides in nuclear medicine treatments. A perspective on the new results in the field of sub-millimeter scale internal dosimetry of tissue and vascular structures will be given, together with a review of the studies on cellular microdosimetry.

1. Introduction

Much of the progress of medical physics would not have been possible without the substantial contribution of information technology. In diagnostic and therapeutic applications of ionizing radiations, proliferating simulation packages would facilitate the operators' tasks at multiple levels. Computer simulations are divided into two major categories, depending on the approach used:

Analytical simulations; into this category falls every simulation adopting an analytical approach, using iterative algorithms that maximize (or minimize) a given function. This process is repeated several times and, according to the values obtained in the previous step, the algorithm calculates the basic parameters (length and direction) of the next step. The advantages of these algorithms are associated with their high computational efficiency and speed of calculation. These benefits are paid, however, in terms of specificity: most of the analytical simulation packages, in fact, are able to consider with approximation only a limited number of physical effects.

Monte Carlo simulations; this category includes all the packages that can address the problem under consideration through a statistical sampling of events, using random number generators. Unlike the case of analytical packages, the main disadvantage of the approach lies precisely in the Monte Carlo calculation time and its related precision (the goodness of the solutions provided increases with the number of simulated events and, consequently, over time). The Monte Carlo approach has, however, greater flexibility and a high popularity that makes its use possible in clinical practice.

Historically, nuclear medicine is the field in which most of the early Monte Carlo calculations in medical radiation physics were performed, and today their use continues to increase. The advent of computer power in hospital departments, mainly in the form of workstations and fast personal computers, has made the Monte Carlo method a common tool in the different areas of research and clinical development where the interaction of ionizing radiation with matter is of importance, namely in x-ray diagnostics, nuclear medicine and radiation therapy, as well as in the radiation-protection aspect in these fields [1].

2. The Monte Carlo Method

The Monte Carlo method describes a very broad area of science, in which many processes, physical systems and phenomena are simulated by statistical methods employing random numbers. The general idea of the Monte Carlo analysis is to create a model, which is as similar as possible to the real physical system of interest, and to create interactions within that system based on known probabilities of occurrence, with a random sampling of probability density functions (PDFs). The Monte Carlo method does not solve an explicit equation, but instead obtains answers by simulating individual particles and recording aspects of their behavior. Numerous particles are followed from their emission by a source, through to their loss by absorption or leakage. The trajectory of each particle is broken down into sequences comprising free flight and, at a given moment, a collision whose nature is randomly selected from a set of possible reactions in the material, the assigned probabilities being related to the cross sections of the material in question. The behavior of real particles within the physical system is predicted from the accumulated data on a large number of simulated particles, and the response of the instrument is determined from the statistical mean of the behavior of the population of the simulated particles. This technique is well suited to solve complicated three-dimensional and time-dependant problems because no averaging approximations are required in space, energy or time. Upon completion of a Monte Carlo simulation, the computational code used gives a statistical uncertainty dependent upon the number of particles that contributed to the result so that the correct convergence of the calculation can be checked. The selection of a statistical error criterion often results from a compromise between calculation time and the required accuracy of the result. But this is not the only contribution to the overall uncertainty of the simulated quantity, and this error does not represent the accuracy of the result compared to the true physical value. A Monte Carlo simulation is limited by the validity of the assumptions and the accuracy of the model used. These limitations come from the level of detail of the geometry model, the accuracy of each input data used during the calculation (material data, nuclear data), the physics treatments and any other interpretational models used to convert calculated quantities into a representation of the instrument response.

The uncertainties arising from the differences between modelling and reality (description simplifications, lack of knowledge of some data, etc.) are estimated using both sensitivity studies, by assessing which relative influence of variation in each parameter on the result, and benchmarking against experience [1].

3. Monte Carlo Packages Employed in Medical Physics

3.1. Geant4

The Geant4 toolkit provides a versatile and comprehensive software package for modern simulation applications that involve the interaction and passage of particles through matter. It can handle complex geometries efficiently and compactly, and allows geometry visualization and particle tracking through a variety of interfaces. It provides simulation for a wide range of physics processes based on theory, data or parameterisations. These treat, for example, hadronic interactions from thermal energies up to 1 PeV; electromagnetic interactions of charged hadrons, ions, leptons and photons from 250 eV to 1 PeV or more; as well as the production and propagation of optical photons. The implementation of the toolkit in an object-oriented design allows it to be easily extended, where necessary, to meet the requirements of the user, through class inheritance. In addition, there are a growing range of utilities for the visualization and analysis of resulting data. The software is freely available at a source-code level over the Web.

Geant4 has been, and continues to be developed and maintained through a Memorandum of Understanding agreed between the many collaborating institutes. This worldwide international collaboration, and the user community, is continuing to expand due to the applicability, accessibility and versatility of this toolkit, with current applications ranging from medical physics to high-energy astrophysics, as well as particle physics and accelerator design.

Among the various therapies using ionizing radiation, metabolic radiotherapy is a widely used technique applied in the treatment of many diseases. It exploits the correlations between the metabolic properties of the organ or tissue to be treated and a particular radionuclide [1, 2].

3.2. FLUKA

FLUKA is a Monte Carlo code able to simulate the transport and interaction of electromagnetic and hadronic particles in any target material over a wide energy range. It is a multi-purpose, multi-particle code that can be applied in many different fields.

Particular attention has been devoted to following the various components of the hadronic and electromagnetic cascades. Thus, the code characteristics at intermediate energies make it particularly reliable in treating problems in the fields of radiotherapy and radiation protection. The code can simulate electromagnetic and μ interactions up to 100 TeV, neutron interaction and transport down to thermal energies (multigroup below 20 MeV) and hadron-hadron and hadron-nucleus interactions up to 100 TeV. Nucleus- nucleus interactions, previously only available above 5 GeV/n (up to 10,000 TeV/n), have been recently implemented also down to ≈100 MeV/n by coupling FLUKA to the Relativistic Quantum Molecular Dynamics code RQMD-2.4. The validity of the physical models implemented in FLUKA has been benchmarked against a variety of experimental data over a wide energy range, from accelerator data to cosmic ray showers in the Earth's atmosphere. FLUKA is widely used for studies related both to basic research and to applications in radiation protection and dosimetry, radiobiology (including radiotherapy) and cosmic ray calculations [1, 3].

3.3. EGS

EGS (Electron-Gamma Shower) is based on the code development in the early 1960s by H.H. Nagel and represents several years of additional development. This code system is a general-purpose package for the Monte Carlo simulation of the coupled transport of electrons and photons in an arbitrary geometry for particles with energies from a few keV up to several TeV. The EGS code system is one of a chain of three codes designed to solve the electromagnetic shower problem by Monte Carlo simulation. This chain makes the simulation of almost any conceivable electron-photon transport problem possible.

The structure of the system, with its global features, modular form, and structured programming, is readily adaptable to virtually any interfacing scheme that is desired on the part of the user. EGS4 is a package of subroutines plus block data with a flexible user interface. EGS4 uses a material cross section and the branching ratio data created and fit by the companion code, PEGS4. EGS4 allows for the implementation of importance sampling and other variance reduction techniques such as leading particle biasing, splitting and path length biasing [1, 4].

3.4. MCNP

MCNP is a Monte Carlo N-Particle simulation package currently used for neutron, photon and electron transport. It is generally employed in radiation protection and dosimetry, radiation shielding, radiography, medical physics, nuclear criticality safety, detector simulation and analysis, and accelerator or reactor designs. The code is able to simulate any three-dimensional configuration of materials in geometric cells. For neutrons, all reactions given in a particular cross-section database (such as ENDF/B-VI) are considered. Thermal neutrons are described by both the free-gas and specific models. For photons, the code includes both incoherent and coherent scatterings, fluorescent emissions after photoelectric absorption, absorbtion in pair productions with local emission of annihilation radiation, and bremsstrahlung. A continuous, slowing-down approximation (CSDA) is used for electron transport that includes positrons, k-shell x-rays, and bremsstrahlung. Important features that enhance MCNP versatility include a powerful general source, criticality source, and surface source, geometry and output tally plotters, several variance reduction techniques and an extensive collection of cross-section databases [1, 5].

4. Validation

The problem of validating Monte Carlo results is particularly important in medical applications, where simulations can be employed to calculate therapeutic dose distributions in often complex geometries, and the inaccuracies related to the calculation method sum up to the inaccuarcies of the experimental data used for the dose quantification. In external beam radiotherapy (EBRT), the main problem is target localization and eventually tracking, but the tissue inhomogeneities near the target volume can also lead to accuracy losses. In nuclear medicine therapies with beta- or alpha-emitting radiopharmaceuticals, the main uncertainties come from the quantification of activity biodistribution in time, carried out through scintigraphic or PET imaging.

Except from hadrontherapy applications, the medical physics applications of Monte Carlo packages involve their electromagnetic interaction simulation capabilities. Geant4 [2] outcomes had been extensively compared with experimental data [6], particularly for electromagnetic interactions [7-9] and in the applicative field of medical physics [10-11].

Over the last few years, new parameterisations were introduced in order to accurately extend electromagnetic interactions down to 250 eV, allowing for the simulation of atomic excitations [8].

These extensions are particularly useful for the dosimetry of low energy electrons such as Conversion Electrons (CE) or Auger electrons, and in the field of microdosimetry [12].

5. Applications in Nuclear Medicine

Historically, nuclear medicine is the field in which most of the early Monte Carlo calculations in medical radiation physics were performed, and today their use continues to increase. The advent of computer power in hospital departments, mainly in the form of workstations and fast personal computers, has made the Monte Carlo method a common tool in different areas of research and clinical development where the interaction of ionizing radiation with matter is of importance, namely in x-ray diagnostics, nuclear medicine and radiation therapy, as well as in the radiation protection aspect in these fields. [1]

An important aspect of the applications of the Monte Carlo technique associated with nuclear medicine is the simplicity of simulating the physics involved in most cases. The energy range of the photons used in this field allows minimum emphasis on the production of secondary charged particles and high-energy processes that yield an electromagnetic cascade; at the other extreme, very low-energy photon interactions can pratically be ignored. This provides a sharp contrast with applications in other areas, such as radiation therapy, were secondary electrons play a major role; or diagnostic x-rays, where photon elastic scattering might be of importance. Calculations of absorbed dose distributions for beta sources obviously fall outside this general characteristic, but low-*Z* materials are involved even in these cases. It is then possible to avoid the extra complication of intricate cross sections for elastic scattering or the treatment of bremsstrahlung processes in detail. On the other hand, detailed simulations of complicated geometry configurations is a field shared today by most other applications, and nuclear medicine is not an exception in adopting the enormous potential provided by some of the major "public" Monte Carlo codes with their advanced geometry capabilities. [1] The applications of Monte Carlo simulation in nuclear medicine can be broadly divided in applications oriented to instrumentation development and optimization, and to the ones oriented to therapeutic dosimetry applications.

If we consider the Geant4 package, a specific toolkit named GATE is available for the former application [13], which allows for the simulation of SPECT or PET tomographs, in order to test and develop new geometries and detector materials in a simulative environment [14-15]. Another powerful and versatile tool for Geant is available for the development of medical applications, GAMOS. [16]

Concerning dosimetric applications, several codes had been extensively employed in order to calculate dose point-kernels [17-18] (i.e., the radial profile of dose around an isotropic, point-like source of radiation). Dose point-kernels are useful in three-dimensional, internal dosiemtry to calculate doses through a convolution process. Another approach to three-dimensioanl dosimetry is associated with the use of dose factors calculated at the voxel level, the so-called voxel S values, introduced by the MIRD commission [19] and currently under study in order to extend their application to different voxel sizes and radionuclides [20-21]. Earlier studies [22] had introduced simplified antropomorphic phantoms and dose factors useful to calculate radiation absorbed doses to each organ due to radioactivity accumulated in the other organs and tissues. Such dose factors were calculated by means of Monte Carlo simulations. Furthermore, specific models and geometries can be studied, and doses from radionuclides can be evaliuated with proper simulations. As an example, we studied the fraction of energy absorbed in ellipsoidal volumes due to the emission of electrons and/or photons within the same volumes [23-26]. These self-dose factors are a useful extension to the commonly employed sphere model, widely employed in nuclear medicine to represent nodules, lesions, or other small uptaking tissues.

Finally, it should be noticed that, with the ever increasing calculation power of computers, the direct Monte Carlo approach became feasible for patient-specific internal dosimetry, as described by some Authors [27].

6. Applications in Radiology

The Monte Carlo simulation has also found widespread applications in diagnostic radiology, particularly in the evaluation of x-ray emission spectra from tubes and for dosimetric studies.

Concerning source spectra, Taleei and Shariari [28] have compared the outcomes from Fluka and MCNP in order to simulate the generation of x-ray spectra between 80 and 120 kV, and the effect of filtration with different absorbing materials, aiming to reproduce experimental data.

Detailed simulations were reported also by Ay et al. [29] and Bontempi et al. [30], who reproduced the typical diagnostic x-ray and mammography spectra, taking into account the possible geometries, materials, and experimental factors such as a ripple factor in high voltages.

These and similar results are very useful both as an input for further dosimetric calculations or simulations, and as a benchmark for quality assurance protocols.

A great effort is spent in the field of computed tomography (CT) simulations, in order to estimate doses resulting from different scan protocols to various antropomorphic phantoms. The difference in scan protocols lies in the type of scan (sequential or spiral), and in the collimation, pitch, kV and mAs employed [31-32]. Specific dose reduction algorithms accounting for the actual tissue depth and attenuation factors are also an interesting subject for study and intercomparison.

The dosimetry of pregnant women undergoing diagnostic CT scans is one of the key applications of Monte Carlo simulations in radiology. A study was developed by Shi et al. [33], using the real CT images from a 30-week pregnant woman as input for a EGS simulation, aimed to determine the doses to organs and tissue of the fetus.

We also studied, through a Geant4 simulation, the dose increase in selected organs during a CT with an administration of an iodinated contrast medium [34], relating it with the increment in CT numbers expressed as Hounsfield Units (HU): we obtained average percent increments ranging from about 20% for liver, to about 70% for kidneys. In this study, the Monte Carlo simulation of simplified antropomorphic neck and abdomen phantoms exposed to a model of a 120-kV collimated source served to quantify the dose increase as a function of the quantity of iodine accumulated in each organ at the time of irradiation, while experimental measurements allowed to relate the iodine concentration to the HU increment, this being the only measurable *in vivo* quantity.

7. Applications in External Beam Radiotherapy

The unwanted dose-increase, in-contrast, enhanced CT can be converted in a therapeutic advantage in the so-called Contrast-Enhanced RadioTherapy (CERT). This effect is maximized for kV photons rather than for MV ones, due to the strong dependence from the atomic number Z of the photoelectric cross-section. The Compton interaction mechanism dominated for MeV photons, and the dose increment due to high-Z iodine accumulated in target tissues was weaker [35-36].

Monte Carlo studies and treatment planning models for CERT continue to be developed for specific applications, addressing some of the critical issues of this therapeutic modality, such as the iodine uptake inhomogeneities within the target tissues [37-38].

A number of studies were concerned with the simulation of medical Linacs. Among the most recent applications, we can cite the new extension of the GATE package to the simulation of diagnostic and therapeutic external sources with Geant4 [39]. Such an extension has been already applied to the simulation of a 6 MV photon beam from a medical Linac [40]. The Monte Carlo simulation is particularly useful in benchmarking conventional treatment planning systems (TPS), and when unconventional treatments and applications are developed. Recently, the possibility to remove flattening filters from IMRT treatment heads was investigated [41] in order to reduce the scattered radiation exposing non-target regions, thus applying the same principle of dedicated IMRT devices such as Tomotherapy or Cyberknife. In order to simulate treatments with these advanced systems, general purpose Monte Carlo codes have been extended in order to account for continuously variable source beams [42] or for complex irradiation geometries [43].

Simulations were applied also to the Gamma Knife device, which uses more than 200 collimated high-intensity ^{60}Co radioactive sources, pointing to a unique central focus to carry out a radiosurgical treatment of brain tumors and vascular malformations. Simulative results from Geant4 were compared with experimental measurements of radiochromic films by Romano et al. [44], while more recently Pourfallah et al. [45] made a similar comparison between EGS outcomes and polymer gel dose meters.

Sealed radioactive sources for brachytherapy were extensively characterized by means of Monte Carlo simulations [46]. Brachytherapy can be conducted with different geometries (interstizial or in contact), with various dose rates and treatment durations, to cure cancers of the cervix, prostate, breast, skin, brain, eye, respiratory or digestive tracts. Several nuclides are employed, such as ^{60}Co, ^{192}Ir, ^{125}I, ^{103}Pd or ^{169}Yb, and Monte Carlo simulations allowed for the dertermination of the dose fields around such sources within realistic models of tissues to be irradiated [47-48].

The last mention must be given to the ever-growing field of hadrontherapy, the external beam radiation therapy exploiting the high-linear energy transfer and the narrow Bragg peak of the high-energy hadron beams. In this advanced field, where accelerators typical of nuclear physics find applications in radiotherapy for the acceleration of protons or heavy ions such as carbon, research sites are designed and treatments are planned through specifically developed Monte Carlo simulations as well [49], such as the ones developed by the INFN project Catana for the proton therapy of ocular melanoma [50].

References

[1] D. Lizio, "Applications of the Geant4 Monte Carlo to Internal Dosimetry of Nuclear Medicine: a New Model for Dosimetry of Ellipsoidal Target", PhD Thesis, University of Messina (Italy) 2009.

[2] Agostinelli S et al. *Nucl. Instrum. Methods A* 506, 250 (2003).

[3] G. Battistoni et al. *AIP Conference Proceeding* 896, 31 (2007).

[4] Kawrakov, *Med. Phys.* 27, 485 (2000).

[5] F. B. Brown et al. ANS Mathematics and Computation Division Topical Meeting, Monterey, CA, April 15, 2007.

[6] K. Amako et al. *Nucl. Phys. B Proc. Suppl.* 150, 44 (2006).

[7] K. Amako et al. *IEEE Trans. Nucl. Sci. Conf. Rec.* 1-7, 2115 (2004).

[8] G.A.P. Cirrone et al. *Nucl. Instrum. and Meth. A* 618, 315 (2010).

[9] Lechner et al. *IEEE Trans. Nucl. Sci.* 56, 398 (2009).

[10] J.-F. Carrier et al. *Med. Phys.* 31, 484 (2004).
[11] B. A. Faddegon et al. *Phys. Med. Biol.* 54, 6151 (2009).
[12] S. Incerti et al. *Med. Phys.* 37, 4692 (2010).
[13] S. Jan et al. *Phys. Med. Biol.* 49, 4543 (2004).
[14] P. Descourt et al. *Phys. Med. Biol.* 55, N253 (2010).
[15] X. Zhu et al. *Phys. Med. Biol.* 54, 4181 (2009).
[16] GAMOS website: http://fismed.ciemat.es/GAMOS/
[17] E. Mainegra-Hing et al. *Med. Phys.* 32, 685 (2005).
[18] L. Maigne et al. *Phys. Med. Biol.* 56, 811 (2011).
[19] W. E. Bolch, et al. *J. Nucl. Med.* 40, 11S (1999).
[20] Dieudonnè, R. F. Hobbs, W. E. Bolch, G. Sgouros, I. Gardin. *J. Nucl. Med.* 51, 1600 (2010).
[21] M. Pacilio et al. *Med. Phys.* 36, 1543 (2009).
[22] W. S. Snyder, M. R. Ford, G. G. Warner, and S. B. Watson (1975). New York, Society of Nuclear Medicine.
[23] E. Amato, D. Lizio and S. Baldari, *Phys. Med. Biol.* 54, 4171 (2009).
[24] E. Amato, D. Lizio and S. Baldari, *Phys. Med. Biol.* 54, N479 (2009).
[25] E. Amato, D. Lizio and S. Baldari, *Phys. Med. Biol.* 56, 357 (2011).
[26] E. Amato, D. Lizio, R. M. Ruggeri, M. Raniolo, A. Campennì, S. Baldari, *Quart. J. Nucl. Med. Mol. Imag.* PMID: 21242948 (2011).
[27] E. E. Furhang et al. *Med. Phys.* 24, 1163 (1997).
[28] R. Taleei et al. *Appl. Radiat. Isot.* 67, 266 (2009).
[29] M. R. Ay et al. *Phys. Med. Biol.* 49, 4897 (2004).
[30] M. Bontempi et al. *Med. Phys.* 37, 4201 (2010).
[31] G. Jarry et al. *Phys. Med. Biol.* 48, 2645 (2003).
[32] Sechopoulos et al. *Radiology* 247, 98 (2008).
[33] C. Shi et al. *Med. Phys.* 31, 2491 (2004).
[34] E. Amato et al. *Med. Phys.* 37, 4249 (2010).
[35] F. Verhaegen et al. *Phys. Med. Biol.* 50, 3555 (2005).
[36] J. L. Robar et al. *Phys. Med. Biol.* 47, 2433 (2002).
[37] H M Garnica-Garza *Phys. Med. Biol.* 54, 5411 (2009).
[38] H M Garnica-Garza *Phys. Med. Biol.* 56, 341 (2011).
[39] S. Jan et al. *Phys. Med. Biol.* 56, 881 (2011).
[40] L. Grevillot et al. *Phys. Med. Biol.* 56, 903 (2011).
[41] M. Delayrd et al. *Phys. Med. Biol.* 55, 7333 (2010).
[42] J. Lobo et al. *Phys. Med. Biol.* 55, 4431 (2010).
[43] M. Constantin et al. *Phys. Med. Biol.* 55, N211 (2010).
[44] F. Romano et al. *IEEE Nucl. Sci. Symp. Conf. Rec.*, 2581 (2007).
[45] T. A. Pourfallah et al. *Med. Phys.* 36, 3002 (2009).
[46] G. Gualdrini and P. Ferrari Eds. ENEA - QUADOS International Workshop Proceedings, (2004).
[47] M. J. Rivard, *Med. Phys.* 38, 306 (2011).
[48] F. J. Casado et al. *Physica Medica* 26, 132 (2010).
[49] T. T. Bohlen et al. *Phys. Med. Biol.* 55, 5833 (2010).
[50] G. A. P. Cirrone et al. *Nucl. Phys. B Proc. Suppl.* 172, 273 (2007).

In: Medical Physics
Editor: Marcin Balcerzyk
ISBN: 978-1-62257-590-9

Chapter VI

Internal Dosimetry in Nuclear Medicine

***Ernesto Amato*[1]*, Domenico Lizio*[2]*, Alfredo Campennì*[1]*, Astrid Herberg*[1] *and Sergio Baldari*[1]**

[1]Department of Radiological Sciences, University of Messina, Italy
[2]Research Center of Saluggia, Institute of Radiological Protection, ENEA, Italy

Abstract

The assessment of radiation absorbed dose is a fundamental issue to plan nuclear medicine therapies with radionuclides. The dose distribution depends on the biokinetics of the radiopharmaceutical and from the physical decay scheme of the radionuclide carried by the molecule. While the physical properties of each nuclide are well known from experimental data, the biodistribution of the radiopharmaceutical within the patient's body depends on the dynamic biologic pathway that, in turn, is governed by the physio-pathologic role of the molecule, by the characteristics of the patient, by the type and stage of the disease, and by the route of administration.

The activity distribution must be sampled several times post-administration, by means of planar or tomographic (SPECT or PET) imaging. Tomographic techniques are rapidly substituting planar whole body imaging, since, thanks also to the accurate attenuation correction and image coregistration brought by a simultaneous CT scan, they reach a spatial resolution and an accuracy in activity quantification unprecedented.

After a general review on the planar and tomographic image quantification techniques, we focus on the dosimetric models, going from the different anthropomorphic models proposed for the whole human body or specific anatomic districts, to 3D techniques based on voxel dose factors, convolution of dose point-kernels and direct Monte Carlo computation. Particular emphasis will be given to the contribution of Monte Carlo simulation to the development of new and more accurate dosimetric and microdosimetric models for internal dosimetry, especially in relationship with the radiobiological models of the effect of the different particulate radiations employed.

We describe the role, specific protocols and obtainable results in the main nuclear medicine therapies such as the 131-I therapy of thyroid diseases, the therapy of neuroendocrine tumors (NET) with somatostatin analogs labeled with beta- or Auger-emitters, and the therapy of non-Hodgkin lymphomas with beta-labeled monoclonal antibodies, focusing on dose-efficacy relationships and on the limiting of side effects to other potentially critical organs.

1. Fundamentals of Internal Dosimetry

Nuclear medicine therapies with radiopharmaceuticals are an effective therapeutic approach for a large number of diseases, ranging from the most classical radiometabolic therapy of hyperthyroidism to the radioimmunologic therapy of lymphomas and the radio-receptorial therapy of neuroendocrine tumors [1-4].

The basic principle of targeted radio-therapy is to deliver to pathologic tissues an amount of radiation dose high enough to destroy them, while preserving other organs and tissues from unnecessary exposure to ionizing radiation.

Such a result is obtained by choosing proper pharmaceuticals whose biodistribution is narrowly targeted on pathologic tissues, and attaching them to a suitably chosen radionuclide. The choice of the best radionuclide is carried on with the aim of maximizing radiation energy deposition in the target tissue during the desired treatment time. Beta-emitters are the best choice in most cases, because beta radiation has a mean range in tissue from few millimeters to few centimeters. Also used are alpha- and Auger-emitters, for millimeter and submillimeter ranges [5-6].\

The decay properties of some of the most widely used electron emitters are listed in Table 1; the maximum penetration range in soft tissue varies between about 10 um for ^{111}In and 1 cm for ^{90}Y.

In order to plan routine and experimental treatments, as well as to verify the effectiveness of an applied treatment, it is fundamental to estimate the response and potential toxicity.

The absorbed dose is the physical quantity whose estimation can give quantitative information about the potential effect of therapy, toxicity, and risk of delayed radiation damages on other tissues.

The quantification of radiation absorbed dose from administered radionuclides is the matter of study of the internal dosimetry [7-8].

Table 1. Average beta and Auger or Conversion Electron energies, and half-lifes of the simulated radionuclides

Radionuclide	$T_{1/2}$ (days)	E_β (keV)	E_{mono} (keV)
^{111}In	2.80	-	2.7
^{177}Lu	6.71	132.9	13.9
^{131}I	8.02	181.5	9.6
^{153}Sm	1.95	226.3	43.9
^{186}Re	3.78	326.1	14.7
^{90}Y	2.67	948.8	0.2

The radiation absorbed dose, commonly intended as dose, is the average energy imparted by the radiation per unit mass:

$$D = \frac{d\overline{E}}{dm} \tag{1}$$

The dose is measured in joules per kilogram or gray (Gy).

A widely employed procedure for the evaluation of the dose absorbed after administration of a radiopharmaceutical was developed by the Medical Internal Radiation Dose (MIRD) commitee (9). The method evaluates the dose to a target organ in two steps: the sources embedded in the target tissue and those in the adjacent tissues.

The dose absorbed by a target k, in the presence of a single source h, can be calculated as:

$$D(r_k \leftarrow r_h) = \widetilde{A}_h S(r_k \leftarrow r_h) \tag{2}$$

where r_k is the target region, r_h indicate the source region, $\widetilde{A}_h$ is the cumulated activity in the source organ and S is the average dose absorbed by the target per unit cumulated activity in the source. The cumulated activity in organ h is the total number of disintegrations in that organ:

$$\widetilde{A}_h = \int_0^\infty A_h(t)dt \ , \tag{3}$$

while the definition of the S factor is:

$$S(r_k \leftarrow r_h) = \frac{\sum_i \Delta_i \varphi_i (r_k \leftarrow r_h)}{m_k} \tag{4}$$

where Δ_i is the average energy emitted per transition as i-th radiation, ϕ_i is the absorbed fraction of such energy by the target r_k emitted by the body source r_h, and m_k is the mass of the target tissue.

For example, in the case of a beta-gamma-emitting radionuclide, the average self-dose, i.e. the dose in case of source coinciding with target volume, is given by:

$$\overline{D} = D_\beta + D_\gamma = \frac{\widetilde{A}\left(\Delta_\beta \varphi_{\beta_i} + \sum_i n_i E_i \varphi_{\gamma_i} \right)}{m} \tag{5}$$

where Δ_β is the average beta energy emitted per disintegration and ϕ_β its absorbed fraction, E_i are the photon energies, n_i and φ_{γ_i} are their emission probabilities and absorbed fractions, respectively.

It can be useful to define the specific absorbed fraction:

$$\Phi_i(r_k \leftarrow r_h) = \frac{\varphi_i(r_k \leftarrow r_h)}{m_k}. \tag{6}$$

Φ has the units of inverse mass.

Thus, the dose D in organ k due to activity in organ h, can be expressed as:

$$D(r_k \leftarrow r_h) = \widetilde{A}_h \sum_i \Delta_i \Phi_i(r_k \leftarrow r_h). \tag{7}$$

In general, if several organs accumulate the radiopharmaceutical, a sum must be carried out over all organs:

$$D(r_k) = \sum_h \widetilde{A}_h \sum_i \Delta_i \Phi_i(r_k \leftarrow r_h) \tag{8}$$

2. Evaluation of Absorbed Fractions

In the calculation of the absorbed fraction, we must take into consideration the irradiation geometry, i.e. the dimension and position of source and target volumes, the nature of the materials traversed by the radiation, and the nature and energy of the same radiations, emitted by the nuclide.

In the following subsections we describe the methods employed to evaluate the absorbed fractions in geometrical frameworks of increasing complexity [9-10].

2.1. Scarcely Penetrating Radiations

The simplest case is for charged particles or photons of very low energy that lose all of their energy after travelling a short distance. If the source volume is much larger than this distance, we can say that the target volume is the same as the source volume:

$$\begin{aligned} \varphi(r_k \leftarrow r_h) &= 0 \quad r_k \neq r_h \\ \varphi(r_k \leftarrow r_h) &= 1 \quad r_k = r_h \end{aligned} \tag{9}$$

2.2. Point source of Monoenergetic Photons in Vacuum

In case of a point source of monoenergetic photons in vacuum, it can be shown that the absorbed fraction in a small amount of substance of area dS, thickness dr, and energy absorption coefficient μ_{en} introduced in vacuum at a distance r is:

$$\varphi = \frac{dE}{\widetilde{A}\Delta_i} = \frac{\mu_{en} drdS}{4\pi r^2} \quad , \tag{10}$$

where $\widetilde{A}\Delta_i$ is the total energy released by the source and dE is the energy absorbed in the small absorber. If the source radiates its energy isotropically, the fraction passing through dS is $dS/(4\pi r^2)$. The fraction of that energy absorbed in dr is μ_{en} dr.

2.3. Point Source of Monoenergetic Photons in an Indefinite Absorber

If the source is not in vacuum but in an infinite, homogeneous and isotropic absorbing medium, the number of photons at distance r from the source is modified by the factor $e^{-\mu r}B(r)$, where the exponential accounts for photon attenuation between source and target points, and the "buildup factor" $B(r)$ accounts for secondary photons produced in the absorbing medium. The absorbed fraction can be calculated as:

$$\varphi = \frac{\mu_{en} drdSe^{-\mu r}}{4\pi r^2} B(r) \tag{11}$$

The buildup factors in water are tabulated for photons of various energies [11].

2.4. Electrons and Photons Uniformly Distributed in an Ellipsoidal Volume

The ellipsoidal shape is often used in nuclear medicine as a model of uniform uptaking tissues or organs, as thyroid nodules or lobes, lymph nodes, or larger organs such as kidneys. In three recent articles, we presented a comprehensive model aimed to calculate the absorbed fraction for electrons and photons characterized by a generic energy spectrum, uniformly distributed within an ellipsoidal volume of soft tissue [12-14]. An application of such method to the dosimetry of hyperthyroidism is presented in [15].

Through a Monte Carlo simulation in Geant4, we obtained the absorbed fraction for different ellipsoidal shapes; oblate, prolate, scalene, and spheres, and we represented them as a function of a dimensional quantity named "generalized radius":

$$\rho = 3\frac{V}{S}, \tag{12}$$

where V is the volume and S the surface of the ellipsoid. The $\varphi(\rho)$ data is represented in Figure 1 for electrons and in Figure 2 for photons, together with the respective fits obtained by means of the analytical form:

$$\varphi(\rho) = \left(1 + \frac{\rho_0}{\rho^s}\right)^{-1}, \tag{13}$$

where the fit parameters ρ_0 and s were in turns fitted with properly chosen functions of the primary electron or photon energy. For electrons we chose:

$$\rho_0(E) = dE^f \exp\left(-gE^{\frac{lnE}{ln^2 10}} \cdot 10^{\frac{h}{\ln E}}\right), \tag{14}$$

and

$$s(E) = \frac{kE+m}{E^r+n} + p, \tag{15}$$

where d, f, g, h, k, m, n, p and r are constants reported in Table 2.

Table 2. Constants for the evaluation of $\rho_0(E)$ and $s(E)$ for electrons [14]

$\rho_0(E)$		$s(E)$	
d	$2.7010 \cdot 10^{-6}$	k	$4.6835 \cdot 10^{2}$
f	1.6181	m	$1.9465 \cdot 10^{3}$
g	$2.3790 \cdot 10^{3}$	n	4.1747
h	7.4563	p	$-4.6656 \cdot 10^{2}$
		r	1.000133

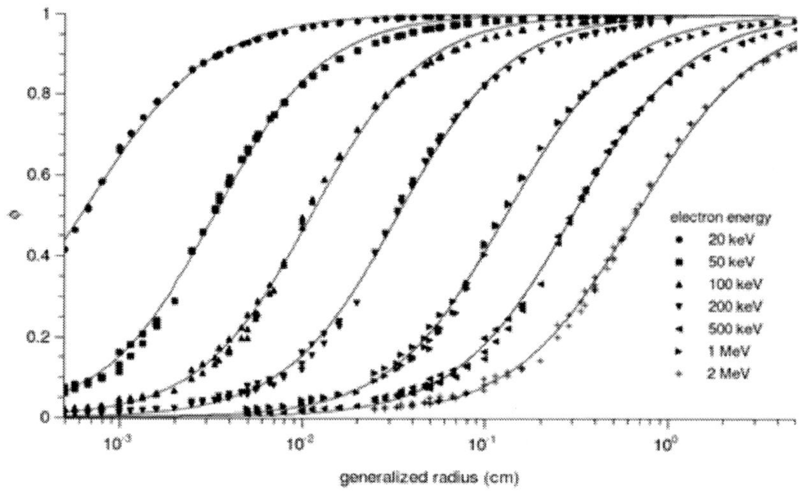

Figure 1. Absorbed fraction for electrons in ellipsoids [14].

Concerning photons, the parameters ρ_0 and s has been fitted with the functions:

$$\rho_0(E)=\frac{gE^m}{1+hE^n\left(1-e^{-E/r}\right)}, \qquad (16)$$

and

$$s(E)=\frac{aE^3+bE^2+E+c}{E^3+dE^2+eE+f}, \qquad (17)$$

where *g, h, r, m, n, a, b, c, d, e* and *f* are constants reported in Table 3.

Table 3. Constants for the evaluation of $\rho_0(E)$ and $s(E)$ for photons [13]

$\rho_0(E)$		$s(E)$	
g	$9.355\cdot10^{-5}$	a	1.118
h	$3.173\cdot10^{-5}$	b	10.036
r	337.25	c	$1.269\cdot10^{5}$
m	3.336	d	-24.823
n	2.947	e	257.722
		f	$1.157\cdot10^{5}$

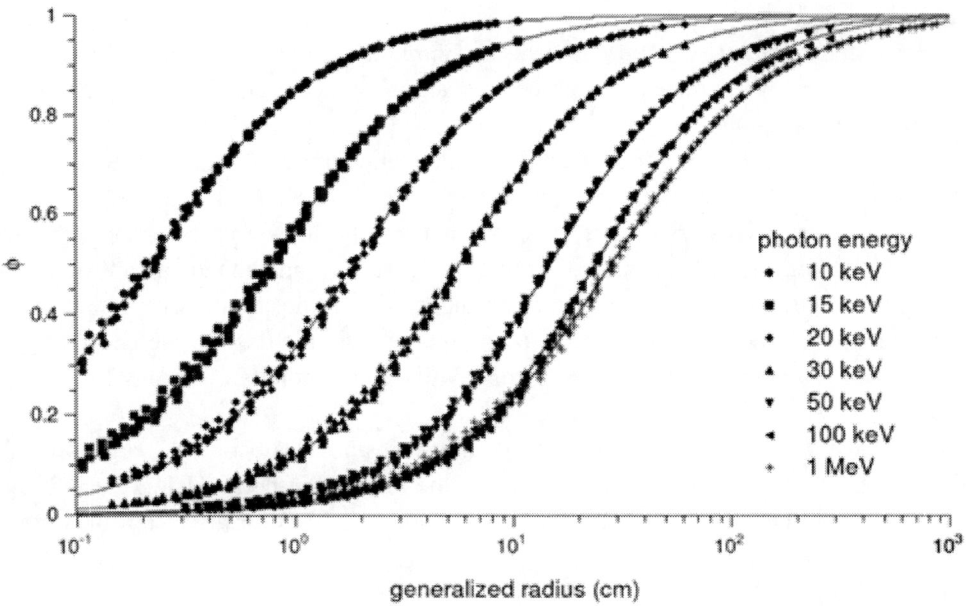

Figure 2. Absorbed fraction for photons in ellipsoids [13].

The dose in the case of an extended beta spectrum will be obtained by integration:

$$D = \frac{\tilde{A}}{m}\int \frac{dn(E)}{dE} E\varphi(E)dE = \frac{\tilde{A}}{m} E_{dep}, \tag{18}$$

where *dn/dE* is the differential emission probability of the radionuclide. The integral represents the average energy deposited per disintegration, E_{dep}.

In the most general case of a radionuclide that emits monoenergetic and beta electrons, and gamma photons, the average dose to the target volume will be given by:

$$\overline{D} = D_\beta + D_e + D_\gamma = \frac{\tilde{A}}{m}\left[\int \frac{dn(E)}{dE} E\varphi(E)dE + \sum_i \left(n_{e,i}E_{e,i}\varphi_{e,i} + n_{\gamma,i}E_{\gamma,i}\varphi_{\gamma,i}\right)\right], \tag{19}$$

where n_e and n_γ are the monoenergetic electron and gamma photon emission probabilities, respectively, of energies E_e and E_γ. φ_e and φ_γ are the electron and photon absorbed fractions.

Such analytical formulation allows to calculate the absorbed fraction and, consequently, the dose for any beta-gamma or Auger-gamma emitter in a generic ellipsoidal shape in the whole range of dimension and ellipticity of practical interest in internal dosimetry.

2.5. Simplified Antropomorphic Models

A widely adopted approach to dosimetry of internal emitters is based on standardized phantoms and specific dose-rates, pre-calculated for couples of Source-Target organs and for various radionuclides.

The advantages of such an approach are the simplicity of measurements and calculations, and standardization of the dosimetric procedure, allowing easier comparisons among results of different research groups.

The limits rely on the simplified shape of mathematical models, that does not account for the real anatomy and for the inter-patient differences.

Tables of average dose *S* factors are published by the MIRD commission [16], which were calculated through Monte Carlo simulation based on mathematical antropomorphic phantoms, in which a uniform distribution of radioactivity in the source organs was assumed. Once the cumulated activity in the source organ *h* is known, the dose to the target organ *k* is calculated with Equation 2, using the *S* factor relative to the couple of organs for the proper radionuclide.

Often, most of the radiopharmaceutical is taken up in one or two organs, and the remaining activity distributes uniformly through the rest of the body. Using the index *h* for the organs with noticeable activity, TB for the total body, and RB for the rest of the body,

$$\tilde{A}_{RB} = \tilde{A}_{TB} - \sum_h \tilde{A}_h \tag{20}$$

The dose to the k-th target organ can be calculated as:

$$D_k = \sum_h \tilde{A}_h S(r_k \leftarrow r_h) + \tilde{A}_{RB} S(r_k \leftarrow RB) \tag{21}$$

The fundamental limits, leading to potentially relevant inaccuracies, are related to the basic assumptions: simplified organ shapes, fixed volumes and reciprocal positions, fixed organ density and chemical composition, each organ is treated as a whole, resulting in average dose calculations.

All of these quantities are strictly patient-specific, as well as the biokinetic distribution of the administered radiopharmaceutical. Moreover, the shape and size of organs can be varied by the disease, and pathologic tissues not assimilable to normal organs can be a target of radiometabolic treatment.

Finally, the treatment of small organ's metastases (e.g. lymphomas) cannot be accurately modelized by simplified phantoms, because each sub-region (e.g. lymph node) is characterized by its proper uptake regime.

3. Three-Dimensional Dosimetry

With the development of emission tomographic imaging modalities such as SPECT or PET, corrected and integrated with co-acquired attenuation maps from a CT scan, the clinical implementation of three-dimensional internal dosimetry techniques became feasible.

In order to obtain a fully patient-specific dosimetry from any radiopharmaceutical administered in radiometabolic therapies, the following is needed: i) the three-dimensional map of the density and composition of the patient's body, together with a definition of the organ geometry; ii) the three-dimensional map of the cumulated radiation activity in the patient's body; iii) the calculation of the volume distribution of the absorbed radiation dose.

The first step can be obtained through a computerized tomography (CT) of the whole body; the 3D map of the tissue density can be derived from the CT Hounsfield Units after an accurate system calibration carried on by means of measurements on phantoms. Semi-automated software for organ contour determination can be employed in order to define organ geometry, while the chemical composition of the organs can be assigned from literature data.

The second information is given by single photon emission computer tomography (SPECT) studies conducted at several times post-administration. During reconstruction, SPECT data must be corrected for attenuation and scattering by means of CT data. SPECT images can be obtained either with γ emission from a β-γ (α-γ or Auger-γ) emitter used for therapy, or with diagnostic doses of the same radiopharmaceutical labeled with a γ-emitter and administered before the therapic dose for treatment planning purposes, or together with the therapic dose for follow-up of treatment.

The possibility to obtain valuable dosimetric information from the bremsstrahlung X-ray emission consequent to the administration of pure β-emitters was also studied.

The third step can be carried out by means of:

1. dose point-kernel convolution methods;
2. voxel *S* value approach;
3. direct Monte Carlo calculation.

The dose point-kernel gives the dose per decay event at a given distance from a point source located inside an infinite and homogeneous absorbing medium. Dose point-kernels for electrons and photons were calculated by means of different Monte Carlo simulations, and are available either in tabular form or as analytical functions obtained fitting the simulative data [17-18].

By convolving the dose point-kernels evaluated for the actual radionuclide spectrum in the three-dimensional distribution obtained from the volume distribution of cumulated activity, obtained from SPECT-CT or PET-CT images, one can retrieve the volume distribution of the absorbed dose. In order to speed up calculations, fast Fourier or fast Hartley transformations can be employed.

The voxel *S* value approach was proposed in 1999 by the MIRD commission [19], and allows the calculation of the volume distribution of the absorbed dose from the three-dimensional map of the cumulated activity and from the knowledge of voxel dose factors, named *S* values, characteristic of the radionuclide and of the voxel size.

Following the MIRD approach, the average dose to the k-th voxel can be calculated as:

$$\overline{D}_k = \sum \widetilde{A}_h \cdot S_{k \leftarrow h}\,, \tag{22}$$

where $\widetilde{A}_h$ is the cumulated activity in the generic h-th voxel and $S_{k \leftarrow h}$ is defined as:

$$S_{k \leftarrow h} = \sum \Delta_i \cdot \frac{\varphi_i(k \leftarrow h)}{m_k}\,, \tag{23}$$

where Δ_i is the mean energy emitted as radiation *i* per decay, φ_i is the absorbed fraction, and m_k is the mass of the *k*-th voxel. In MIRD Pamphlet 17, voxel *S* values are tabulated for five radionuclides (^{32}P, ^{89}Sr, ^{90}Y, ^{99m}Tc and ^{131}I) and two cubic voxel sizes (3 and 6 mm). Recently, Dieudonnè et al. [20] presented a generalization of such an approach to variable voxel sizes, which uses a resampling procedure to obtain *S* values for a generic voxel size of practical interest for nine radionuclides, while Pacilio et al. [21] had pointed out that different Monte Carlo codes can lead to differences in the calculation of voxel *S* values.

Finally, the Monte Carlo approach is the most demanding in terms of computational power, but it is the only one which can account accurately for tissue inhomogeneities. When the three-dimensional map of tissue density and composition from CT scan and the volume distribution of the cumulated activity through SPECT or PET emission tomography are known, it is possible to generate a fixed number of decay events statistically distributed in space according to emission tomography data within the patient's body reconstructed from CT and track all of the emitted secondaries in order to calculate the geometrical distribution of energy deposition, i.e. the three-dimensional map of absorbed dose, and finally to rescale the results to the actual values of cumulated activity [22-23].

4. Radiopharmaceuticals Biokinetics Measurements

4.1. Kinetics of Radioactive Decays

Let us consider a radioactive sample containing, at the time t, N radioatoms of the same nuclide [9].

We know that the rate of change of N, i.e. the activity $A(t)$ representing the number of decays per second, is proportional to N through the decay constant λ:

$$A(t) = \frac{dN}{dt} = -\lambda N \tag{24}$$

The integration of Equation 24 leads to the well known exponential decay law for the number of radioatoms in a radioactive sample:

$$N = N_0 e^{-\lambda t} \tag{25}$$

The decay constant λ is related to the decay time τ and to the half-life $T_{1/2}$ by:

$$T_{1/2} = \frac{\ln 2}{\lambda} = \tau \ln 2 \tag{26}$$

The SI unit of activity is the *becquerel* (Bq), defined as 1 transition per second:

$$1Bq = 1 transition \cdot s^{-1}, \tag{27}$$

while the old unit of activity was the *curie* (Ci):

$$1Ci = 3.7 \times 10^{10} Bq \tag{28}$$

We said that the cumulated activity $\widetilde{A}_h$, defined in Equation 3, is the total number of transitions in in the organ h; it is measured in $Bq\ s$, which is dimensionless. The ratio between the cumulated activity in h and the administered activity A_0 is named residence time:

$$\tau_h = \frac{\widetilde{A}_h}{A_0} \tag{29}$$

Even if τ has the physical dimensions of a time, it must not confused with the decay time of a radionuclide.

In fact, the residence time is the length of time that activity at a constant rate A_0 would have to reside in the organ to give that cumulated activity.

In the following subsections, we will discuss how to calculate cumulated activities and residence times in cases of increasing complexity.

4.2. Immediate Uptake with No Biological Excretion

In this simplified model, a fraction F_h of the radiopharmaceutical is accumulated very rapidly in the organ h, and it remains there. The cumulated activity will be:

$$\tilde{A}_h = A_h \int_0^\infty e^{-\lambda t} dt = \frac{A_h}{\lambda} = \frac{F_h A_0}{\lambda} \tag{30}$$

The residence time will be:

$$\tau_h = \frac{A_h}{A_0} = \frac{F_h}{\lambda} = \tau F_h \tag{31}$$

4.3. Immediate Uptake with Exponential Biological Excretion

If the radiopharmaceutical, once instantaneously accumulated in an organ, is washed out from it at an exponential rate, the variation of N with time is governed by an effective decay constant:

$$\frac{dN}{dt} = \lambda_{eff} N \qquad \text{where} \quad \lambda_{eff} = \lambda + \lambda_{bio} \tag{32}$$

where the biological decay constant is characteristic of the biological washout process in act. In this case, the number of radioatoms in the organ h follows the monoexponential decay law:

$$N_h(t) = N_0^h e^{-\lambda_{eff} t} \tag{33}$$

and the cumulated activity will be:

$$\tilde{A}_h = A_h \int_0^\infty e^{-\lambda_{eff} t} dt = \frac{A_h}{\lambda_{eff}} \tag{34}$$

4.4. General Case

In real cases, the uptake of a radiopharmaceutical in an organ or tissue can require a certain amount of time and, thus, the assumption of instantaneous uptake can lose in accuracy, leading to an overestimation of the cumulated activity.

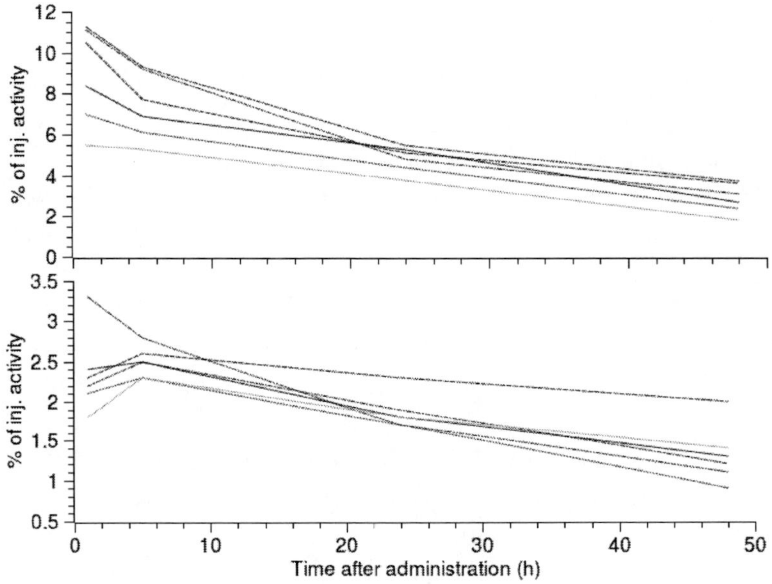

Figure 3. Uptake curves for kidneys (upper panel) and spleen (lower panel) during ^{111}In-DTPA-Octreotide dosimetric scans, for six patients.

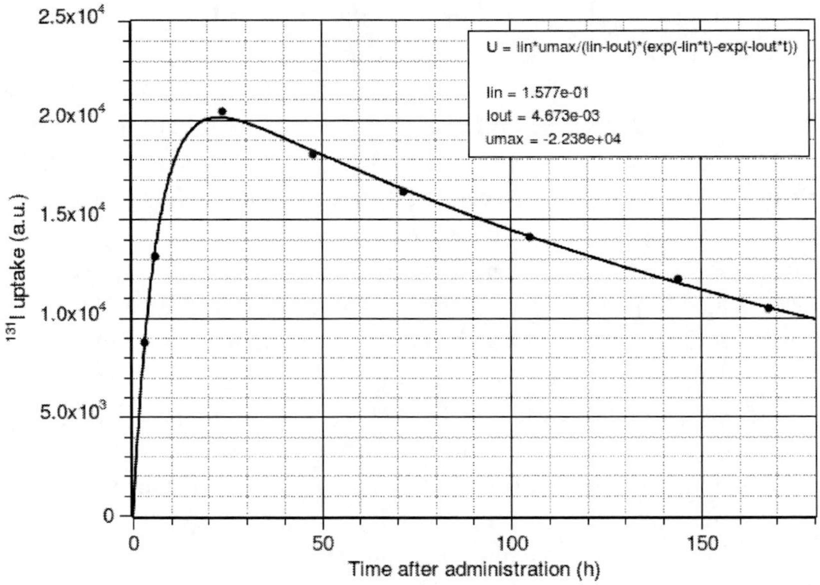

Figure 4. ^{131}I uptake curve acquired by means of a scintillation probe at eight times after oral administration; solid line represents the analytical fit with the indicated function U=$U(t)$.

On the other hand, the washout phase can not be accurately described by a simple monoexponential decay; often, a bi-exponential curve fits better to a biokinetical behavior composed by a first phase of rapid clearance in which the biologic half-life is smaller than the physical half-life, followed by a slower retention phase in which, on the contrary, it is the physical half-life that governs the overall effective half-life.

As examples, in Figure 3, we show the uptake curves for kidneys and spleen during diagnostic ^{111}In-DTPA-Octreotide whole body scans for six patients. Octreotide is a somatostatin analogue, used for the diagnosis of neuroendocrine tumors. Acquisitions were taken at 1, 6, 24 and 48 hours post-injection. We can see that the biokinetics in kidneys can be represented by an instantaneous uptake followed by a bi-exponential washout, while for the spleen, the maximum uptake is at 6 hours for 5 of the 6 patients.

In Figure 4, we present a typical ^{131}I uptake curve for a hypethyroid patient, acquired by means of a scintillation probe at eight times after oral administration of a diagnostic activity (1.8 MBq). The maximum radioiodine uptake is around 24 hours, the uptake phase is described by an exponential growing characterized by an uptake constant λ_{in}, while the washout phase is governed by a monoexponential decay with an effective decay constant λ_{out}.

5. Data Acquisition and Image Quantification Techniques

The quantification of the activity taken by organs and tissues as a function of time allows, as said above, to calculate the cumulated activity, accomplishing thus to the first step for dose calculations. Usually, data is acquired through in-vivo measurements of uptake carried out by means of simple probes or nuclear medicine imaging devices. This data can be completed by radioactivity measurements of biological samples of urine or blood.

Non-imaging in vivo measurements are currently employed mainly for thyroid dosimetry, and are carried out using scintillation or solid state detectors equipped with a cylindric lead collimator. Such a detector, connected with an electronic acquisition chain, allows energy discrimination of incident gamma rays and thus permits the quantification of unscattered radiation emerging from the source organ in the field of view of the collimator.

Well counters employing similar detectors with energy discrimination capability are used for activity measurements in biological samples.

By far the most complex procedure is needed in order to obtain quantitative data from imaging devices, such as gamma cameras for planar scintigraphy and SPECT (Single Photon Emission Computed Tomography), or PET (Positron Emission Tomography).

When using planar scintigraphic imaging, the most accurate image quantification method is the conjugate view approach, described in the MIRD Pamphlet n. 16 [24]. In this case, static or whole-body antero-posterior (A/P) images are taken with a dual head gamma camera equipped with parallel hole collimators, maintaining a constant distance of the camera heads with respect to the patient's body.

A region of interest (ROI) is drawn around the uptaking organ considered and the measured intensities I_A and I_P are used to calculate the radionuclide activity with the formula:

$$A=\sqrt{\frac{I_A I_P}{e^{-\mu_e t}}}\frac{f_j}{C} \qquad (35)$$

where C is the calibration factor of the gamma camera, expressed as count rate per unit activity of the radionuclide considered, t is the overall thickness of the patient's body in the measurement point, and μ_e and f_j are, respectively, the effective linear attenuation coefficient and a correction factor accounting for the source thickness:

$$\mu_e=\frac{1}{t}\sum\mu_i t_i \quad \text{and} \quad f_j=\frac{\mu_j t_j/2}{\sinh(\mu_j t_j/2)} \qquad (36)$$

Here, μ_i and t_i are the linear attenuation coefficient and the thickness of the i-th layer of tissue lying below or above the source organ j.

Such formulation holds only in cases of negligible activity in the superimposed tissues in the background. For the general case, a more complicated calculation is needed, taking into account also the activity in the superimposed regions and/or in the background.

In order to overcome the inherent limits of planar imaging, such as organ superimposition, SPECT or PET tomographic images are gaining diffusion in internal dosimetry, especially when combined with CT images for accurate attenuation correction.

When co-registered CT images are not available, ordinary attenuation correction methods such as the one developed by Chang are employed. Furthermore, in order to correct for Compton scattering within the patient, proper correction methods can be implemented such as dual or triple energy window acquisitions, methods based on convolution and deconvolution techniques or spectral fitting.

6. ^{131}I Therapy of Hyperthyroidism

Radio-iodine treatment (RIT) is a well established method for the therapy of hyperthyroidism due to Graves' disease (GD), toxic adenoma (TA) or toxic multinodular goiter (TMG). The optimal ^{131}I activity to be administered is still a matter of debate. Several studies have evaluated the effectiveness of different dosimetric methodologies, with variable and often controversial results, in order to the frequency of recurrence and hypothyroidism. [25-26].

Thus, the ^{131}I activity for RIT can be calculated by a dosimetric approach, or uncalculated (standard approach). In both cases it is necessary, if possible, to suspend anti-thyroid drugs treatment 4-6 days before RIT or thyroid scintigraphy.

In the standard approach, the ^{131}I activity for RIT is usually about 555 MBq. In these patients, the scintigraphy with radioiodine thyroid uptake can be not necessary.

In the dosimetric approach, it is necessary to make some diagnostic procedures: a) thyroid ultrasonography to calculate the volume of the whole gland, or the hot thyroid nodule(s), assuming for them an ellipsoidal shape; b) thyroid scintigraphy and radioiodine thyroid uptake (RTU) measurements after ^{131}I tracer activity administration (1.8 MBq).

The dosimetric method proposed by the MIRD commission [16] is the most employed; in this approach, RTU measurements are usually conducted between 3 and 120 hours after administration. Multiple RTU measurements permits to obtain the effective half-life of the radioiodine into hot nodules or whole gland (Figure 4).

In TMG and TA patients, it is useful to calculate the effective volume of the hot nodules, taking into account the presence of involving area(s) (whole nodule volume - involving area volume).

In general, the radiation doses to be imparted to the gland (GD) or to the nodule(s) (TA and TMG) are up to 200 Gy and in the range 200-300 Gy, respectively.

In GD with ophtalmopathy, the radiation dose to be imparted must be properly adjusted. In these patients, it is useful to administer anti-inflammatory drugs for at least four weeks after the therapeutic radioiodine administration.

In all patients, the choice of the radioiodine therapeutic dose will be determined also in function of: sex and age, type of hyperthyroidism (overt or sub-clinical) and its temporal duration, concomitant Hashimoto thyroiditis, anti-thyroid drugs therapy and its temporal duration.

Basing on the above considerations, the activity to be administered can be calculated as:

$$A = 5.829 \frac{Dm}{U_{max} T_{1/2eff}} \qquad (37)$$

where A is the activity to be administered in MBq in order to impart a dose D in cGy to the target (nodule or gland) of mass m in g, U_{max} is the maximum per cent radioiodine uptake (usually measured at 24 hours) and $T_{1/2\ eff}$ is the effective half-time in hours.

More recently, Amato et al. [15] proposed an improved method to calculate the activity to be administered during radioiodine treatment of hyperthyroidism. In this approach, a Monte Carlo simulation had been employed to retrieve the radiation dose in nodules or in the whole gland within an antropomorphic phantom, taking into account the ellipsoidal shape of the target volumes.

As a result, the activity to be administered can be calculated with the formula:

$$A = \frac{Dm}{U_{max} T_{1/2eff}} \cdot \frac{32.31\,\rho + 1}{\rho(0.2625\,\rho + 5.1819)} \qquad (38)$$

where we recall the definition of generalized radius $\rho = 3V/S$, stated in Equation 12, and here expressed in cm. The volume and the surface of the ellipsoidal target (nodule or thyroid lobe) can be calculated from the semiaxes measured through ultrasonography as:

$$V = \frac{4}{3}\pi abc \qquad S = 4\pi \left(\frac{(ab)^p + (bc)^p + (ac)^p}{3} \right)^{1/p} \qquad (39)$$

7. ^{131}I therapy of Differentiated Thyroid Carcinoma

Total or Near-Total thyroidectomy (TT or NTT) followed by radioiodine, ^{131}I, ablation of thyroid remnant (TRA), is considered the best approach for the treatment of differentiated thyroid carcinoma (DTC). Indeed, numerous authors, such as Mazzaferri et al [27], have demonstrated that the prognosis quod vitam and/or survival curve significantly improve if RIT follows TT or NTT.

The TRA permits a better management of the follow-up of these patients, because in the absence of thyroid tissue, the serum thyroglobulin (Tg) shall be undetectable. Thus, every enhancement of Tg serum can be considered as a relapse of disease. The RIT can be carried out in hypothyroidism (TSH>=30) or under exogenous TSH stimulation (rhTSH). The post-dose whole-body scan, carried out 4-8 days after RIT, permits to identify thyroid remnant and methastases (loco-regional and/or distant).

For TRA, there is not wide agreement for the choice of the ^{131}I activity to be administered. The TRA activity can be standard or calculated by a dosimetric approach [28]. In the standard case, the ablation activity most employed is 3700 MBq, less frequently, lower activities of 1110 or 2220 MBq are employed. The dosimetric approach should require the pre-therapeutic scintigraphic scan with ^{131}I at diagnostic activity or ^{124}I PET; the activities calculated by dosimetry not differ significantly from the standard values. Thus, in the clinical practice, the dosimetric approach is not employed frequently.

On the contrary, the dosimetric approach is useful for the treatment of methastases, particularly in cases of lung and/or bone methastases, where standard activities can lead to a radiation dose imparted to the lesions lower than necessary. For the dosimetric approach, it is necessary to acquire some whole scans after radio-iodine therapy. Whole body scans are acquired from 6 hours to 7 days post-RIT. [29-30]

8. Peptide Radio-Receptorial Therapy of Neuroendocrine Tumours

Over the last decade, Peptide Receptor Radionuclide Therapy (PRRT) has become an important tool for the treatment of metastasized and/or inoperable neuroendocrine tumors (NETs). This is an heterogeneous group of epithelial neoplasms originating from neuroendocrine cells embriologically derived from the neural crest (NE). They occur as a part of familial syndromes (MEN) or sporadically and had a substantial increasing rate of incidence and prevalence in the last 30 years, partly due to improvements in diagnostic imaging which lead to the incidental diagnosis of asymptomatic cases.

NETs are potentially ubiquitary and most frequent sites include the gastro-entero-pancreatic tract followed by the lungs (less frequent localizations include the skin, the adrenal glands, the thyroid and the genital tract). They present different biological behaviors: some may be aggressive and highly malignant (poorly differentiated NETs) but the great majority are either benign or relatively slow-growing (well differentiated NETs) and can retain many differentiation features as the possession of neuroamine uptake mechanisms, the expression of specific markers (NSE, Cromogranin A) and the hormone secretion associated with an inherited syndrome. Another typical feature of differentiated NETs is the expression of several specific receptors on the cell membrane. Most endocrine GEP tumors possess somatostatin receptors (SSTR) and can therefore be visualized by somatostatin receptor imaging with radiolabeled somatostatin analogue (^{111}In-OCTREOSCAN, ^{68}Ga-DOTANOC). These tracers bind with high affinity to subtypes 2, 5 e 3 of SSTR and provide a relative approximation of somatostatin receptor density, resulting in a positive predictive value of the therapeutic efficacy of somatostatin analogues, based on excellent correlation between in vitro and in vivo receptor evaluation.

On this basis, peptide receptor radionuclide therapy (PRRT) was developed using similar somatostatin analogs with different therapeutic radionuclides.

^{111}In-DTPA-octreotide (OctreoScan), the same radiopharmaceutical used for imaging, has been administered for therapeutic purposes in pilot trials and in some limited experiences. Besides the gamma radiation that is used for scintigraphic scans (γ photons of 172 and 245 KeV), ^{111}In also emits Auger and conversion electrons with a medium-to-short tissue penetration range (0.02-10 and 200-500 μm, respectively). ^{111}In-DTPA-octreotide indeed showed a therapeutic effect dependent on octreotide-receptor complex internalization (by endocytosis) and its traslocation to nuclear compartment, where short path length Auger or conversion electrons are able to reach the target (DNA). Clinical PRRT trials with high doses of ^{111}In-DTPA-octreotide showed promising therapeutic effects but partial responses were seldom achieved: the most accredited explanation is that ^{111}In electrons fail to reach DNA helix, possibly due to the lack of nuclear receptors that were never definitely demonstrated [31].

Better results (objective response in about 30%) are obtained with somatostatine analogs labeled with beta-emitting radioisotopes; among these, ^{90}Y-DOTATOC and ^{177}Lu-DOTATATE are preferred for their advantageous physical properties. [32].

^{90}Y (E_{max} = 2.28 MeV, R_{max} = 11 mm, $\langle E \rangle$ = 0.935 MeV; half-life 2.67 days) has a pure Beta-emission with long range particles that, besides the direct action, lead to irradiation also of no-receptor expressing tumor-cell (cross-fire) which is preferred especially for treatment of larger and inhomogeneous lesions.

^{177}Lu has a longer half life (6.7 days) and lower energy beta-emission (E_{max} = 0.497 MeV; R_{max} = 2 mm) that allows to concentrate all of the energy inside smaller tumors; in addition, ^{177}Lu emits gamma-radiation (113 keV and 208 keV) suitable for scintigraphy and dosimetry during PRRT.

To date, the therapy administration protocols rely essentially on empirical criteria with injection of standard activities, as derived from escalation studies or clinical experience, at variable time intervals (6-12 weeks).

Dosimetric analysis provides evaluation of biodistribution, pharmacokinetic, health organ and tumor absorbed doses and biological effective doses, representing a precious guide for therapy optimization and prevention of toxicity. The data needs to be assessed specifically for the single patient and for the radiopeptide used, as the peptide influences the biokinetics.

The gamma-emission of ^{111}In and ^{177}Lu-peptide enable imaging, dosimetry and therapy with the same compound. In the case of ^{90}Y-peptide, for the lack of gamma emission, the same analogue radiolabeled with ^{111}In or positron emitters as ^{86}Y or ^{68}Ga can be used as surrogates for dosimetric imaging. Furthermore, some recent studies have been published for

^{90}Y-peptide dosimetry with bremsstrahlung images through SPECT-TC systems, or exploiting the rare positron decay of ^{90}Y imaged by new generation PET-TC scanners.

Data collection consists of acquisition of multiple planar whole-body (WB) scan to derive biokinetics over the time (at least 4-5 acquisitions at different times), complementary SPECT imaging to quantify intra-organ activity distribution (especially at the level of the kidneys), WB transmission imaging for attenuation correction, blood and urine samples. Anatomical CT or ultrasonographic data are important for organ mass evaluation [33].

Dose calculation can be performed in the framework of MIRD formalism. A dedicated software (OLINDA/EXM) provides average dose estimates at the organ level assuming standard phantom models, with possible organ mass adjustment according to patient data. Voxel dosimetry method or direct Monte Carlo simulation can provide more reliable dose estimates.

Dosimetry results show common characteristics in pharmacokinetics of radiopeptides: they have fast blood clearance and urinary elimination, the hottest organs are the spleen, the kidneys and the liver and no uptake is visible in the bone marrow in normal conditions.

The adsorbed doses, especially in regards to the tumor, are affected by wide intra-patient variation between treatment cycles.

The adsorbed dose to the tumor is higher during the first two cycles with a gradually reduced uptake in the successive treatments, probably due to saturation or down-regulation of peptide receptors. Tumor adsorbed dose should exceed 100 Gy to obtain therapeutic effect with better results by higher doses (~230 Gy).

For ^{90}Y-peptide and ^{177}Lu-peptide, the kidneys are dose-limiting organs. Due to their physical characteristics, for the same administered activity, the adsorbed doses with ^{90}Y-peptides are higher than those with ^{177}Lu-peptides. The kidney adsorbed dose is reduced consistently (25-65%) by infusion of protective agents and by adopting hyper-fractionated treatment protocols.

The dose to kidneys should not exceed a limit value of 28 Gy, established for external beam radiotherapy, which results in a 50% probability of developing severe late renal failure. Nevertheless, for the difference between the two radiation therapy modalities, this recommendation has been questioned and the biological effective dose (BED) is considered a more accurate quantity, with a good correlation with loss of kidney function.

BED for a given organ or tissue is defined as:

$$BED = \sum D_i + \frac{\beta}{\alpha} \frac{T_{rep}}{T_{rep} + T_{eff}} \sum D_i^2 \tag{40}$$

where D_i is the dose imparted by the i-th treatment cycle, T_{eff} the effective half-life of the radiopharmaceutical in the organ, α is the intrinsic radiosensitivity, β is the repair capacity and T_{rep} is the repair time of the sub-lethal damage. For kidneys, it is usually assumed for $\alpha/\beta = 2.4$ Gy and for T_{rep} a value of 2.8 hours.

Data in literature shows that patients with risk factors, such as hypertension and diabetes, should not receive a BED higher than 28 Gy, while patients with no risk factors might have a renal BED threshold of 40 Gy [34].

References

[1] Paes, F.M., et al. *Semin. Nucl. Med.* 40, 89 (2010).
[2] Goldsmith, S.J. *Semin. Nucl. Med.* 40, 122 (2010).
[3] Pool, S.E., et al. *Semin. Nucl. Med.* 40, 209 (2010).
[4] Sarkar, S.D. *Semin. Nucl. Med.* 36, 185 (2006).
[5] Kassis *Semin. Nucl. Med.* 38, 358 (2008).
[6] Neves, M., et al. *Appl. Rad. Isotop.* 57, 657 (2002).
[7] Zanzonico, P.B. *J. Nucl. Med.* 41, 297 (2000).
[8] Sgouros, G. *J. Nucl. Med.* 46, 18S (2005).
[9] Lizio, D. "Applications of the Geant4 Monte Carlo to Internal Dosimetry of Nuclear Medicine: a New Model for Dosimetry of Ellipsoidal Target", PhD Thesis, University of Messina (Italy) 2009.
[10] Yoriyaz, H., et al. *Med. Phys.* 36, 5198 (2009).
[11] Loevinger, R., Budinger, T.F., Watson, E.E. *MIRD Primer for Absorbed Dose Calculations*, Society of Nuclear Medicine, New York, (1988).
[12] Amato, E., Lizio, D., Baldari, S. *Phys. Med. Biol.* 54, 4171 (2009).
[13] Amato, E.,Lizio, D., Baldari, S. *Phys. Med. Biol.* 54, N479 (2009).
[14] Amato, E., Lizio, D., Baldari, S. *Phys. Med. Biol.* 56, 357 (2011).
[15] Amato,E., Lizio, D., Ruggeri, R.M., Raniolo, M., Campennì, A., Baldari, S. *Quart. J. Nucl. Med. Mol. Imag.* EPUB 18-01-2011 PMID: 21242948 (2011).
[16] Snyder, W.S., Ford, M.R., Warner, G.G., Watson, S.B. *Absorbed Dose per Unit Cumulated Activity for Selected Radionuclides and Organs. MIRD Pamphlet 11,* Society of Nuclear Medicine, New York, (1975).
[17] Mainegra-Hing,E., et al. *Med. Phys.* 32, 685 (2005).
[18] Maigne, L., et al. *Phys. Med. Biol.* 56, 811 (2011).
[19] Bolch, W.E., Bouchet, L.G., Robertson, J.S., *et al. J. Nucl. Med.* 40, 11S (1999).
[20] Dieudonnè, Hobbs, R.F., Bolch, W.E., Sgouros, G., Gardin, I. *J. Nucl. Med.* 51, 1600 (2010).
[21] Pacilio, M., et al. *Med. Phys.* 36, 1543 (2009).
[22] Furhang, E.E., et al. *Med. Phys.* 24, 1163 (1997).
[23] Furhang, E.E., et al. *Med. Phys.* 23, 1523 (1996).
[24] Siegel,J.A., et al. *J. Nucl. Med.* 40, 37S (1999).
[25] Regalbuto, C., et al. *J. Endocrinol. Invest.* 32, 134 (2009).
[26] Giovannella, *Radiol. Med.*, 105, 12 (2000).
[27] Mazzaferri, E.L. *Thyroid* 7, 265 (1997).
[28] Lassmann, M. *Dosimetry Concepts of radioiodine therapy for the treatment of differentiated thyroid cancer,* EANM Dosimetry Committee
[29] Eschmann, S.M., et al. *Eur. J. Nucl. Med. Mol. Imaging* 29, 760 (2002).
[30] Sgouros, G., et al. *J. Nucl. Med.* 45, 1366 (2004).
[31] Kwekkeboom, D.J., et al. *J. Nucl. Med.* 46, 62S (2005).
[32] Cremonesi, M., et al. *Quart. J. Nucl. Med.* 54, 37 (2010).
[33] Cremonesi, M., et al. *Quart. J. Nucl. Med.* 55, 155 (2011).
[34] Bodei, L., et al. *Eur. J. Nucl. Med.* 35, 1847 (2008).

In: Medical Physics
Editor: Marcin Balcerzyk
ISBN: 978-1-62257-590-9

Chapter VII

Optimization in High Precision Radiotherapy

Ludwig Bogner*
Department of Radiotherapy
University of Regensburg, Germany

Abstract

In the last decade, radiotherapy of tumors changed considerably towards high precision. Regarding the process of radiotherapy as a chain, starting from 3D-diagnostics via treatment planning towards irradiation of the patient, each link must be precise in order to achieve an overall accuracy, which is needed in order to improve the treatment outcome. The outcome can be described in terms of cure rate of the tumor at a simultaneously acceptable low normal tissue complication rate.

One challenge is the precision of treatment planning algorithms, which are of outstanding importance to insure uncomplicated cure of the tumor disease. Another development of the last years concerns the optimization of the dose distribution. Compared to by-trial-and-error methods, more and more automatic optimization algorithms are applied. This method is termed *intensity modulated radiotherapy* and it is based on so-called inverse algorithms. Inverse algorithms need a precise dose calculation code, as otherwise they themselveswould introduce an additional error. Dose calculation by Monte-Carlo (MC) simulation is most precise, therefore, inverse MC optimization is shown to overcome these problems.

Nowadays, high precision radiotherapy has to take care of movements of target and normal tissue between different treatments of a fractionated irradiation scheme as well as of intra-fractional movements. Attempts to overcome these problems can be summarized by the term *adaptive radiotherapy.* This term can be even extended by *biologically adaptive radiotherapy*. It regards new molecular imaging methods like e.g. positron emission tomography or magnetic resonance spectroscopy. By these techniques, 3D-images of certain biological information of the tumor may be achieved, as e.g. the

* Tel.: ++49 (0)941 9447602; Ludwig.bogner@klinik.uni-regensburg.de

proliferation rate of tumor cells or the state of hypoxia, which influences outcome severely. It is described how this information can be introduced into biological adaptive radiotherapy.

High precision radiotherapy can be optimized by the methods described above, but of course, intrinsic physical restrictions cannot be overcome. The dose distribution of high energy photons, which are applied extensively in radiotherapy, shows a nearly exponential decrease with depth within the tissue. In contrast, heavy charged particles like protons behave completely different, because their interaction with tissue leads to a finite range. This leads to a so-called inverse depth dose curve with a low dose at the skin surface, a dose maximum near the range and no dose beyond. This Braggpeak, as it is called, leads to an ideal treatment modality. Especially when applied in a scanning mode of pencil beams, intensity modulated proton therapy is a very precise technique, by which the dose to normal tissue may be reduced by more than a factor of 2. It is shown that inverse MC optimization is applicable to protons as well as to photons combined with remarkable error reduction. Moreover, the use of MC simulation for proton dose calculation reduces critical errors in regions with pronounced heterogeneous tissue densities. A technique is described, that cares additionally for the exclusion of beam directions that suffer more from dose uncertainty than others.

In conclusion, methods are described that may reduce to a great extent errors and uncertainties in modern optimized radiotherapy with photons and protons.

1. Introduction

Radiation therapy of malignant tumors is based on the effect of ionizing radiation on biological tissues. This effect was detected soon after the discovery of ionizing radiation more than 100 years ago. Nowadays, mostly Bremsstrahlung-photons with very high energies in the range of 4-21 MV (accelerating voltage) are produced in electron linear accelerators and directed towards the patient.

As tumors in general are of a highly non-regular shape and the radiation doses needed to control them are very high (typically in the range of 60-70 Gy), special care has to be put on healthy tissues of the body. This is especially true for those organs with a higher risk of damage by radiation. The effect of dose to tumor tissue and normal tissues, respectively, is described by the terms tumor control probability (TCP) and normal tissue complication probability (NTCP). Obviously it is desirable to irradiate the tumor such that a maximum complication-free cure, termed P_+ (Schultheiss and Brahme in [1]) is achieved.

$$P_+ = TCP \cdot (1 - \prod_k NTCP_k) \qquad \text{eq. (1)}$$

The index k is denoting different normal tissues. A high value of an uncomplicated cure is only achievable by sophisticated treatment techniques with many beam directions and a 3D- treatment planning system, which allows a precise simulation of the dose distribution prior to the radiation application. The 3D-dose distribution should be as conformal to the tumor shape as possible. Dose calculation is herewith based on computed tomography imaging (CT). After dose calculation, the planning system allows the evaluation of treatment plans including comparisons of different plans, such enabling quantitative ranking of plan qualities. Very often there is a conflict situation between the dose needed for tumor eradication on one hand and the healthy tissue toxicity on the other hand, therefore, the dose calculation has to be very precise. But the dose calculation is not the only issue which has to be precise, every link of the whole chain beginning from diagnostic imaging via the treatment planning system and the so-called verify and record system (v&r) to the linear accelerator has to be precise as well. The v&r-system is a data base system which gets all data necessary to transfer machine parameters for a distinct treatment to the accelerator and which additionally controls the data handling and patient scheduling process, increasing the patient's safety dramatically. The accelerator has to be properly calibrated in terms of radiation dose and controlled by an extended quality assurance program. Such a procedure is termed 3D-conformal radiotherapy (3DCRT) and it can be visualized as a chain, containing all steps within the process of radiotherapy of a patient. The links of the chain are

- Indication for a radiotherapy treatment
- Patient fixation and positioning, followed by a virtual simulation
- CT for treatment planning
- Target definition, prescription of dose and fractionation
- 3D treatment planning, dose optimization and verification
- Radiation treatment
- Treatment verification and control

Despite the high precision, which can be achieved within each link of the chain, obviously some imprecision remains in every link leading to an overall imprecision and uncertainty. By that, further dose escalation to the tumor is not achievable per se even it would be necessary from the point of view of the biological demand to increase the TCP. Therefore, many attempts have been made in order to minimize imprecision in each individual link. It is the purpose of this chapter to show up in detail the sources of imprecision and uncertainty and the attempts to meet them. Some of the presented techniques are in the stage of clinical implementation, others are in a scientific research state.

2. Improving Dose Precision and Reducing Geometrical Uncertainty in Radiotherapy

2.1. Algorithms for 3D-Dose Calculation and Its Precision

In the 1980's, computer workstations were introduced in radiation treatment planning. Based on a computer tomography investigation of the patient, dose distributions could be calculated in at least one CT-image slice, in which the central beam axis was positioned, or optionally in some off-axis slices. The information of each pixel of the CT-images is by definition expressed as so-called Hounsfield units (HU) as a relative measure of the local linear attenuation coefficient $HU = 1000 \cdot (\frac{\mu - \mu_w}{\mu_w})$, where the index w denotes water. HUs are then translated into local electron density values for each voxel by an appropriate calibration. As dose calculation algorithms matrix methods were first applied, which make use of tables derived from dosimetric data, which were measured in water. Later, at the end of the 1980's, separation methods were introduced into treatment planning. These analytical methods make use of theories on the separation of dose contributions from direct and scattered particles [2]. As the model makes use of data from measurements, which are performed in pure water, appropriate corrections according to the heterogeneous density of human tissues have to be performed. Several approximations have to be introduced additionally in order to care for e.g. oblique beam incidence on the skin surface, different source-to-skin distances and field sizes as well as irregular shaped field apertures. Thus, a relative good estimate of the 3D dose distribution could be achieved with the separation method as long as calculations were not performed in areas of the body with pronounced tissue density heterogeneities, differing from the density of water.

The calculation of remarkably improved 3D dose distributions in PC-workstation based systems started in the early 1990's by Ahnesjö [3] and others. They proposed and developed new fluence-based dose algorithms, known as superposition/convolution systems. The radiation transport is described in these models by the energy fluence Ψ_0 of the radiation source, which is the number of transported particles per cm^2 times the energy transported by the particle to a distinct point. The energy released per unit fluence at a distinct point within the patient can be described by a quantity termed, TERMA (T, total energy released per mass), which is proportional to the linear mass attenuation coefficient $T = \psi_0 \cdot \frac{\mu}{\rho}$. The transport of the locally released energy is described by point kernels h. After integration over all points within the patient, where primary radiation is transported, one gets for the dose in an arbitray point within the patient. As all of these quantities are depending on energy, integration over the energy spectrum of the radiation source has to be accomplished additionally.

$$D(\vec{r}) = \int_V \int_E d^3\vec{r}\,'dE \cdot T(\vec{r}\,', E) \cdot h(\vec{r} - \vec{r}\,') \qquad \text{eq. (2)}$$

The integral was solved by numerical methods after discretization of the problem. The great challenge at these times was to establish a method for the transport of the released energy to any point of the patient, as described by the kernel h. It was the development of Monte-Carlo (MC) simulation and one of its early application to solve radiation transport problems that pushed development of superposition/convolution systems. Thus, after MC-simulation of the spread of the energy released in a point or alternatively in a pencil beam-like line through the patient for monoenergetic radiation analytic modeling of these simulated 3D-distributions allowed the determination of dose spread all over the patient's volume.

Primarily the kernels are derived from a set of MC-simulations with different monoenergetic radiation impinging on a semi-infinite slab of water to adapt the non-homogenous distribution of the accelerators output. The heterogeneity of the tissue is regarded by an appropriate density scaling of the kernel in different directions. So-called pencil beam kernels apply distinct approximations such as rotational symmetry for the scatter contribution, such ending up in certain dose inaccuracies especially in regions with distorted secondary electron equilibrium. This occurs in areas with pronounced density gradients or the build-up region. Under these assumptions one can make use of the convolution theorem in order to accomplish fast integration by fourier transformation. This very powerful convolution method had to apply additional approximations in order to compensate for deficiencies such as kernel-tilting because of the beam divergence or oblique incidence on skin surface. But it has to be stressed, that the development of convolution methods such as the so-called pencil-beam algorithm [3] was a great success for clinical treatment planning, because it allowed for the first time sufficiently accurate dose calculation within reliable time, at least in tissue areas without pronounced density heterogeneity.

Many investigations have been performed in areas of the body, where the approximations lead to pronounced errors. These areas are, for example, the chest with the lung and the mediastinum or the head-and- neck environment, other areas with high-gradient density interfaces or close to the body surface. To account better for these problems, superposition methods have been developed to overcome disadvantages of the convolution methods. By applying an angular dependent point spread kernel, the approximations in dose determination in heterogeneous areas are much less serious. Density scaling is performed in these methods not averaged over 2π but only averaged into the secondary particle transport direction within cones. Methods, such as the collapsed cone algorithm [3] diminishes many of the errors being known from pencil beam algorithms, but at the price of a much longer calculation time.

The question arises, why MC-simulation cannot be applied in dose calculation from first principles instead of only applying it to kernel determination within fluence-based methods. The reason for that is the extremely long simulation time, which would be necessary to achieve a dose accuracy with a standard deviation of 2%, which is necessary to end up with an overall uncertainty of about 7-8% in radiation treatment. Nevertheless, some sites started in the late 1990's with MC-simulation codes in radiotherapy dose calculation. An available code was EGS4 [4] among others, developed at the National Research Council Canada (NRC) [5], which was precise and generally applicable (e.g. to nearly all media and many particles and processes). But EGS4 was not dedicated to radiotherapy application with X-rays and because of its general applicability extremely slow. Nevertheless, time was up to introduce MC-simulation into radiotherapy planning. Therefore, some groups applied codes like EGS4 by furnishing the simulation engine with a setup part for an easy input of the patients data like the CT-study, the internal structures (tumor, normal tissues) and the beam setup. Additional attempts have been made, for example, by Fippel [6] to overcome difficulties by developing new MC-codes which are dedicated to the radiotherapy task by restricting the applicability of the code to X-rays and their secondaries and to media representative for biological tissues. The codes were mostly benchmarked by EGS4 or later on by its improved version EGSnrc [7]. BEAM was developed at NRC [8] as a code, especially designed for the simulation of the complicated treatment head of medical linear accelerators. They allow for exact geometrical definition of the constituents of the head with its correct material specification. The simulation results in a phase space file for all simulated particles upstream of the parts of the treatment head, which are indivual for a patient or even a subfield of his treatment plan. This helps a lot in understanding the action and its influence of the different devices of the linear accelerator (linac) head on the very complicated radiation field distal of its exit and allows for improvements. The next step is to trace each particle from the phase space file through the patient's individual parts of the linac-head, e.g. the field-forming multi-leaf collimator (MLC) and then the patient (regarded as a phantom) itself. Among the clinically most succesfull MC codes is XVMC, which is dedicated to radiotherapy and therefore fast, but nevertheless precise. As an analytical model to simulate the treatment head of a linac, the algorithm VEFM [9] is applied, which is similar to analytic head models of superposition/convolution algorithms [3]. Many comparisons have been performed between conventional and MC dose algorithms in different sites of the tumor within the body to evaluate and compare the errors, made by the conventional codes. It could be shown by Jeraj et al. [10] that, for example, in tissue interfaces near the lung, dose errors of up to 8% may occur. Meanwhile, MC codes enter commercial treatment planning systems and, extrapolating the performance of modern PC workstations, a splendid future can be predicted for MC dose engines. An excellent overview over clinical applications of MC codes is given in a report of the American Association of Physicists in Medicine by Chetty et al. [11].

2.2. Treatment Planning: From 'by-Trial-and-Error' to Inverse Planning

Inverse Planning and Intensity Modulated Radiotherapy

A remarkable progress came up in radiation therapy with the introduction of fluence-based 3D-dose algorithms, because dose distributions could now be calculated in the whole body volume. Dose calculation was possible at a high accuracy level, at least in areas of the body, where the condition of secondary electron equilibrium is fulfilled in a good approximation. This condition is considerably violated in the chest or head-and-neck region but not in other areas of the body. With the availability of superposition/convolution algorithms much more complicated treatment techniques were developed, consisting of many irregular beams from anatomically adapted directions, off-axis fields (where the beam axis is outside the field) or combinations of fixed beams with off-axis rotational fields and electron fields. All these techniques aim to achieve a dose escalation within the tumor and simultaneously to lower the dose to healthy tissues, especially to the organs at risk. This technique is termed 3D conformal radiotherapy (3DCRT). Isodose plans are evaluated by means of dose-volume-histograms (DVH) for the structures originally defined, as e.g. the planning target volume and the different organs at risk. The DVHs are evaluated in an integral form to express the dose distributions within each defined structure by proper integration over the differential DVHs.

3DCRT-planning is usually practiced in clinical routine by or under the supervision of an experienced medical physicist, who designs some more or less suitable beam setups for the case under consideration. Subsequently an evaluation of the different dose distributions is performed together with the physician to make a decision. This is a typical 'by-trial-and error' procedure. In the early 1990's it was realized that treatment planning is a typical optimization problem, where all parameters influencing the irradiation can be a target for optimization, as to be the fluence distribution of all beams, the beam and the collimator angles, the beam quality or energy respectively as well. First attempts to solve this so-called inverse problem of radiotherapy by fluence optimization were suggested by Bortfeld et al. [12]. He discretized beams into beamlets and the patient volume into voxels and introduced weighting-factors for each beamlet of all beams. Such the fluence can be modulated by changing the set of beamlet weights. By introducing these weights into the expression for TERMA. the dose distribution in the patient can easily be calculated for any modulated fluence by:

$$T = w \cdot \psi_0 \cdot \frac{\mu}{\rho} \qquad \text{eq. (3)}$$

The inverse problem was solved by an iterative optimization algorithm, mostly by steepest decent gradient methods. As an objective (or scoring, or cost) function f for the optimization physical criteria, like a quadratic dose deviation from a prescribed dose distribution all over the patient's volume, or at least the VOIs was applied. Mostly dose-based quadratic objective functions are applied.

$$f = p^{PTV} \cdot \sum_{i=1}^{N^{PTV}} (D_i - D_0^{PTV})^2 + \sum_{k=1}^{N_k} p_k^{OAR} \cdot \sum_{i=1}^{N_k^{OAR}} (D_i - D_0^{OAR_k})^2 \quad \text{eq. (4)}$$

where D_0 denotes the prescribed dose value, PTV the planning target volume, OAR the organs-at-risk, N is the number of voxels in the relevant structure and p is a penalty factor to allow for relative importance weighting of the different volumes of interest (VOI). Here, the second term is designed for serial organs-at-risk. These are tissues like the spinal cord with clear dose threshold, above which severe damage occurs. The so-called parallel organs-at-risk such as the lung present a pronounced volume effect. This means that the threshold depends strongly on the treated volume of the organ being very sensitive in case of large treated volumes and less sensitive in case of small volumes. These tissues are handled in a different way in the objective function *f* by evaluating the dose statistics (expressed as integral dose-volume-histogram, described below) after each iteration step to allow for the determination of the volume that gets a dose above the threshold. A penalty is applied in such a dose-volume objective if the calculated volume exceeds the tolerated volume. The iterative process of the gradient method is finished after the score value fulfills some minimum criteria, like a re-increase during iteration. Objectives can be defined as physical hard or soft constraints or as dose volume objectives in the DVH. Hard constraints force the optimizer to strictly regard certain dose limits, whereas soft constraints permit a violation of limits, but exert penalization proportional to that penalization. There is an ideal situation, termed infinitron by Schulz [13], characterized by a homogeneous high dose to the target and no dose to the normal tissues, which should be aspired but can never be reached by the optimizer (Figure 1).

Penalty factors are applied to each term within equation (4) for the different VOIs to allow for proper steering of the iterative optimization process with regard to relative importance weighting or individual penalization of the different VOIs. One important constraint in inverse planning is the so-called non-negativity constraint, which has to exclude any negative fluence, by which the optimizer would otherwise try to remove dose from a sub-volume from the treatment plan. Overviews over inverse planning algorithms are given by Bortfeld [12, 14]. Fluence optimization by inverse planning normally leads to improved theoretical dose distributions, which are clinically not realizable. The reason for that is that the fluence discretization into beamlets is performed with a high spatial resolution, which is physically not applicable by the linear accelerator. Therefore, another step called segmentation or sequencing is necessary to translate each fluence modulated beam into a sequence of different subfields termed *field segments* to be shaped by the leafs of the MLC. The apertures and the individual weight of each segment of a beam are such, that the optimal fluence distribution is approximated as good as possible. Detailed studies show that a loss of plan quality comes along with the sequencing. Different approaches have been investigated in order to improve plan quality by introducing fluence smoothing terms into the objective function [15] or by re-optimizing the segment weights or/and shapes after sequencing [16].

As a result of sequencing, the shapes of all segments of the beams are available together with the corresponding relative weights of the segments. The optimized beam setup has to be recalculated by a MC-simulation of all segments in another step, termed *final dose calculation*. The superiority of the re-optimization procedure over fluence smoothing was shown by Rickhey et al. [16]. In Figure 1, the loss of plan quality after sequencing the optimized fluence is demonstrated on a lung cancer case. Additionally a remarkable recovery of the plan quality is observed after re-optimization [17]. The whole procedure is applied as fluence or intensity modulated radiotherapy (IMRT). IMRT was introduced in the early 2000's, but spread slowly in clinical application.

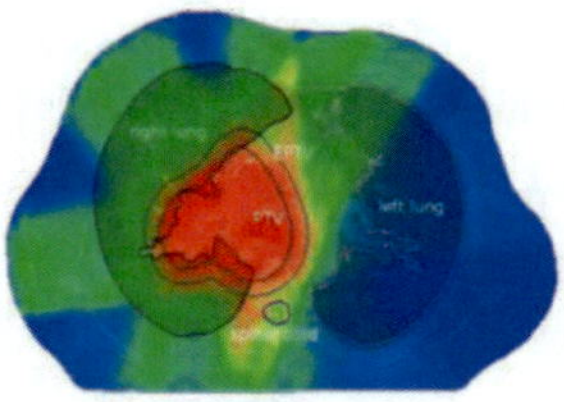

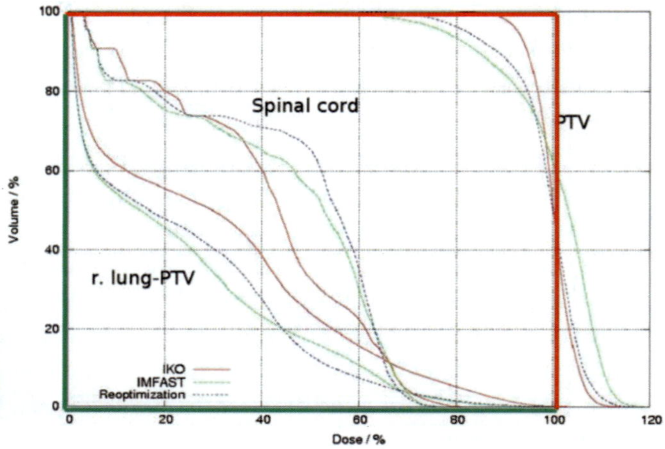

Figure 1. Top: Transversal CT-slice with the dose distribution of a lung cancer case, treated with 5 photon beams. Bottom: Integral dose-volume-histogram (DVH) The hypothetical ideal situation termed infinitron [13] to be aimed for is shown, where 100% of the target volume gains 100% of dose (red) and no dose to all organs at risk (green) is applied. The realistic results of a fluence optimization by inverse kernel optimization performed with IKO [17] for the lung cancer case are shown for the target and the two organs at risk, the spinal cord and right lung minus tumor target volume (PTV). A considerable aggravation by the segmentation process (performed by IMFAST, Siemens) and an improvement of plan quality by the subsequent re-optimization step can be observed.

The reason was, that typical treatment plans consist of 50-70 small segments instead of 5-10 fields in 3DCRT. The problem with that is that any plausibility checking gets lost. Therefore, forced by legal regulations, a 3Ddose verification has to be performed individually for each patient's plan. This is a rather time-consuming effort.

Dosimetric Uncertainties: Consequences from Dose Engine Approximations

Another problem was fixed to be serious in areas where pronounced dose calculation errors may occur, as in cases of strongly heterogeneous tissue environments like for head-and- neck or lung tumors. Whenever the dose algorithm determines an incorrect dose in a subvolume of the body originating from any approximations, then the optimizer tries to correct this deviation by an erroneous fluence modulation in order to maintain the local prescribed dose. Such not only the systematic dose error but additionally the induced convergence error, as it was termed by Jeraj et al. [10] is introduced. They find total errors up to 8% in selected clinical cases.

Inverse Monte-Carlo Optimization

As any deterministic or stochastic change of parameters, which determines the fluence distribution of the beams enforces a recalculation of the 3D-dose distribution, inverse planning is an iterative procedure. This is an extremely time-consuming act if MC-simulation is applied instead of superposition-/convolution-methods. Therefore, only laboratory versions have been developed for comparative investigations of treatment plans for the purpose of quantification and separation of errors [28]. Another concept [18] applies superposition/convolution dose algorithms for optimization and includes an integrated refinement of the fluence based on MC-simulation.

A new paradigma is to regard MC-simulation as a computer experiment by tracking several millions of primary and secondary particles traversing the patient. From that Bogner et al. [19] derived, that feedback controlling should be a possible way to regulate the optimization process according to the dose and dose-volume objectives during the computer experiment. Inverse MC Optimization (IMCO) is such a system that repeatedly applies criteria to control the dose distribution during optimization [20].

From these dose evaluations during the simulation process, each after a distinct number of simulated particles, a change in fluence modulation can be derived from the evaluation of the deviation between the calculated and the prescribed dose distribution. Thus, the system is driven towards a fluence modulation of the beams by which a dosimetric optimum near the prescribed dose distribution can be achieved.

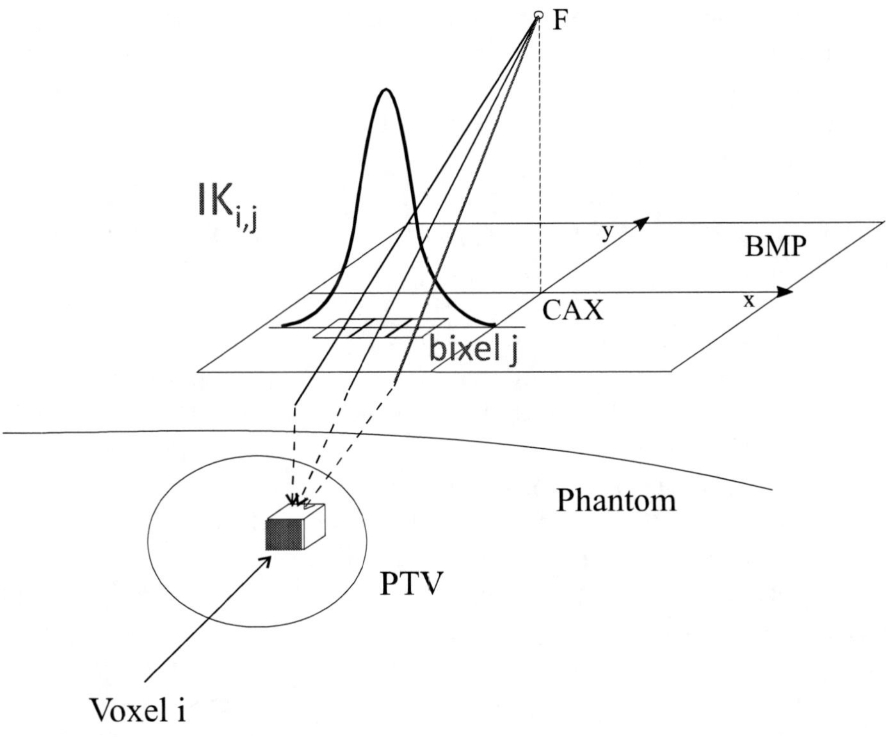

Figure 2. Schematic view of the inverse kernel (IK). IK describes the contributions of all beam elements of a beam termed bixel, discretized in the beam modulation plane (BMP) to a distinct voxel of the patient's volume. F (focal point). CAX (central axis). PTV (planning target volume).

The fluence modulation is derived by use of the, as it was termed by Bogner et al. [17,19] inverse kernel matrix (IK_{ij}) which quantifies the contributions of a distinct bixel number j (the fluence of a narrow beam, discretized in the so-called beam-modulation plane) to any voxel number i of the patient's volume (Figure 2).

The complete IK-matrix can be derived during one single MC-simulation prior to the optimization procedure, in which the history (to denote the bixel number j) is stored on dose deposition in the voxel with number i. For each voxel of the patient's volume, an individual matrix is needed and therefore a memory problem arose in IMCO. This problem can be solved by storing the kernel information in a sparse matrix technology for data compression [20].

A paradigma change occurred in abstaining the special normalization of the inverse kernels introduced in IMCO. Unnormalized IKs can directly be applied to 3D-dose calculation with MC-precision.

$$D_i = \sum_j w_j \cdot IK_{ij} \qquad \text{eq. (5)}$$

This concept, termed *inverse kernel optimization* (IKO) allows a separation of the inverse MC problem into two parts: (i) A pre-step MC-simulation, during which the complete set of IK-elements is derived and stored as a lookup-table. This part is relatively slow, but introduces the MC-accuracy into inverse planning, (ii) another iterative (either deterministic or stochastic) optimization step to optimize the beams's fluence modulations according to any objective function. This part is very fast, as IK-elements are available in lookup-tables from the pre-step simulation.

This type of separation fits optimal to clinical practice, where normally several optimization steps are accomplished interactively in order to adapt the dose-volume-objectives properly. Similar concepts have been introduced in the literature termed dose deposition coefficients for inverse treatment planning with superposition /convolution or MC dose algorithms [21,22].

The Segmentation Problem

Electron linear accelerators are only able to apply radiation fields with a homogeneous fluence over the aperture of a distinct beam (often termed a *field*). The question arises, how the modulated fluences resulting from inverse planning can be delivered. Nowadays, the standard method is either static or dynamic sequencing of each beam, the latter of which is not in the scope of this chapter. In the static mode, each modulated field is sequenced into a number of subfields, termed *segments*, which in superposition yield a good approximation of the modulated field. The sequencing is done by algorithms, which are developed individually for distinct MLCs, having several options for optimization according to treatment time, number of segments, minimal segment shape and size or other influence parameters. At the end of sequencing, a set of segments for each beam and the corresponding segment weights are available. As a consequence of the fluence sequencing into segments, a loss in plan quality is correlated.

The quality of an isodose plan can be expressed in terms of values from parametrization of the DVH, as e.g. $H=D_5\text{-}D_{95}/D_{av}$ for the tumor or V_X for a OAR, where X is a dose value in

% of the reference dose defined in the target. There are three approaches to improve plan quality after sequencing within the IMRT procedure:

1. Introduction of a smoothing term into the objective function [15,16]. By that, smoother fluences result from optimization which diminish the resulting number of segments and lowers the loss in plan quality by sequencing.
2. The introduction of smoothing terms leaves a non-quadratic objective function, which therefore enforces to apply stochastic instead of deterministic search engines to assure finding the global optimum instead of any local minimum. To avoid exhaustive stochastic searches, the sequencing process is amended by a re-optimization of the fluence weights. For that, another MC-simulation of all segments is performed and the dose contributions of each segment are stored separately. Their weighted sum yields the total dose, but the segment weights are subject to another optimization according to

$$f = \sum_i (\sum_s w_s \cdot D_{s,i} - D_{i,0})^2 \qquad \text{eq. (6)}$$

Figure 1 shows a comparison of results of such an optimization procedure, clearly indicating the loss of plan quality after sequencing and partial re-establishing of the quality after re-optimization.

3. Two issues motivated to develop another approach in order to avoid plan quality loss during optimization procedure. Introducing linear accelerator machine parameters directly as parameters for optimization instead of virtual fluences, the concept of a direct-aperture-optimization DAO was developed Shepard et al. [23]. Another strong argument is that seqencers are not available for any new development of MLCs and there is no need for such a sequencer in DAO, as will be shown below.

2.3. Optimizing Treatment Planning by Direct MC Optimization

DAO

The DAO approach was first suggested and demonstrated by Shepard et al. [23]. They derived dose deposition coefficients by MC-simulation with BEAM and EGS4, and applied simulated annealing (SA)[24,25] to optimize a dose-based objective function. Similar systems have been proposed by Bergmann et al. [22] and Bogner et al. [26]. Bush et al. [27] performed a MC-simulation of the treatment head of the accelerator with open fields and they collect simulated particle parameters (type, position, direction, energy) in a phase-space-file (psf), which refers to the position at the top of the MLC. All particles of the psf are sorted into discretized bixels, then followed by a MC-simulation of each individual bixel within the patient to yield the dose distributions of the narrow beamlets. From these dose distributions, the dose deposition coefficients can be derived and stored for the optimization procedure. The following optimization procedure is essentially identical to Shepard's DAO. In contradiction, Bogner et al. [26] developed a method termed 'direct MCOptimization' (DMCO), which applies the inverse kernel matrix derived from one single MC-simulation with the MC code XVMC.

Direct Monte-Carlo Optimization

In DMCO, the beam setup is amended by a pre-definition of the aspired number of segments per beam. All segments are by default regarded to be identical and shaped according to the target in beams-eye-view mode (tumor view from beam direction) plus an additional margin to assure all beamlets hit the target.

The SA procedure starts by sampling a parameter from the list of parameters to be optimized. This list contains all leaf positions from all segments of the beam setup and additionally, all segments weights. After that, the position of a leaf (if the choice was a leaf; if not, then of a segments weight) is changed by another sampling. The sampling of a new leaf position is weighted by Gaussians with the maximum at the former leaf position.

By that, leaf jumps to nearer positions are more advantageous over long-distance-jumps. If the choice was a segment weight, a new weight is sampled. After sampling of a new parameter value, the objective function has to be recalculated (Figure 3). One big advantage of the IK-concept is that changes in dose distribution can simply be performed by adding or subtracting the corresponding bixel contributions accordingly.

Another advantage is the simplicity of checking, if the new leaf-position can be realized and does not violate technically allowed MLC-constraints. The new configuration is accepted or disclaimed according to the rules of SA until a solution at or at least near the global optimum is found.

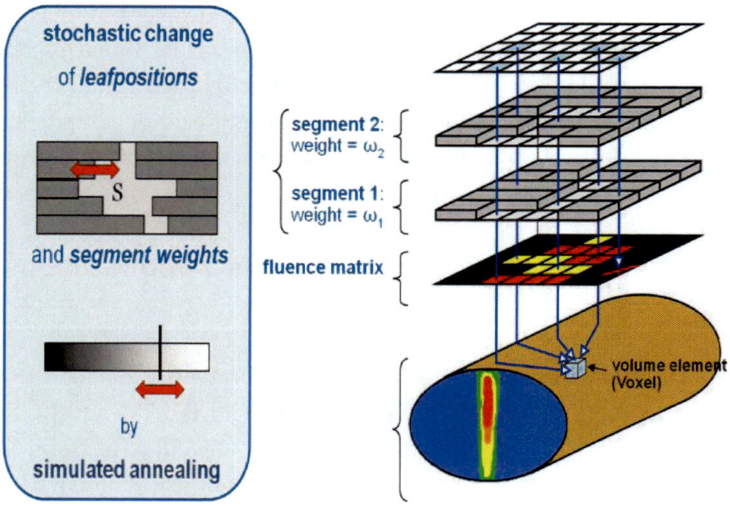

Figure 3. Principles of direct Monte-Carlo optimization (DMCO). Left: Leaf positions and segment weights are stochastically changed and optimized by simulated annealing technique. Right: Composition of the fluence matrix by individual segments and its dose contribution to a distinct voxel of the tumor target.

As a result, the beam fluences are immediately available in terms of segment shapes and their weights, such making sequencing and re-optimization dispensible. The 3D-dose distribution can be calculated very fast from the IK-dose engine by summing only over bixels from inside the segments apertures.

$$D_i = \sum_{k \in segment} w_k \cdot \sum_{j \in bixel(k)} \Theta_{k,j} \cdot IK_{i,j} \qquad \text{eq. (7)}$$

If necessary, a final dose calculation can be applied by a direct MC-simulation of all segments individually. However, investigations of phantom and patient cases revealed that this step may be dispensable because of the precision of the IK dose engine [26,28]. The work flow of DMCO is shown schematically in Figure 4.

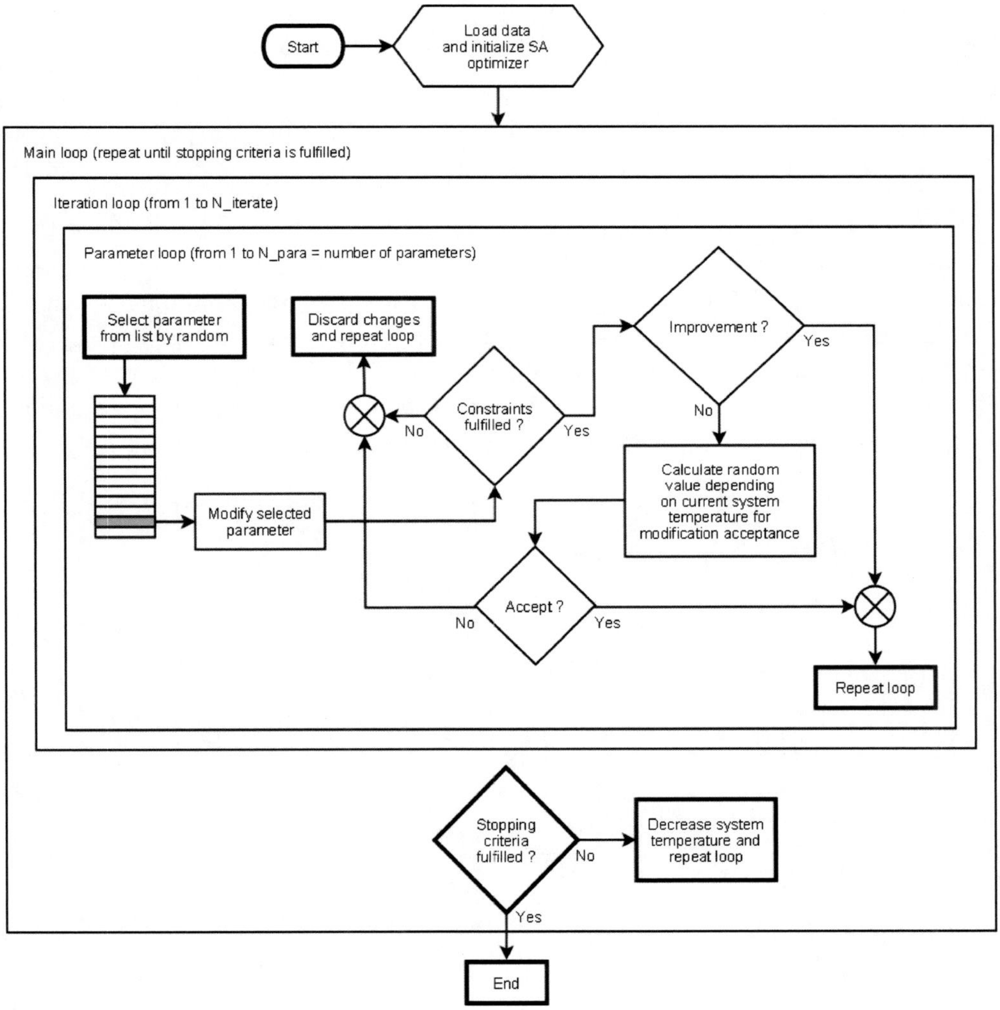

Figure 4. Workflow of the process of direct Monte Carlo Optimization (DMCO) [53, 26].

From Physical to Biological Objective Functions – From Deterministic to Stochastic Optimization

Treatment planning is, even nowadays, mostly optimized according to physical objectives, either dose or dose-volume based. Any optimization of the biological effect of a 3D-dose distribution is, if at all, performed manually by a subsequent evaluation according to biological models like the equivalent uniform dose (EUD) [29], tumor control probability (TCP) and normal tissue complication probability (NTCP) or the complication free cure rate (P+), mostly based on the linear-quadratic (LQ) model (Schultheiss and Brahme in [1]).

Physical optimization is mostly applied due to the lack of reliable LQ-parameters for tumors and normal tissues from clinical evidence-based studies. But as molecular biology methods increasingly enter radiotherapy, the amount of reliable data also increases. A publication of Ling et al. [30] posted a future of optimization in radiotherapy according to biological target view instead of beams eye view by including different biological information into the process of treatment planning. Niemierko suggested a quantity termed EUD, which is defined for any heterogeneous dose distribution over the target as a mean dose influencing the same biological effect as the heterogeneous dose distribution on the tumor tissue. Later, the concept was generalized to any tissue including normal tissue by the gEUD model [31],

$$gEUD_{voi} = \tfrac{1}{N}(\sum_{i=1}^{N} D_i^a)^{\frac{1}{a}} \qquad \text{eq. (8)}$$

where N is the number of voxels in the VOI and *a* is a parameter influencing the biological behavior of the tissue. Therefore, the parameter *a* has to be fixed according to empirical knowledge about the considered organ. There are several approaches with gEUD-based objectives in optimization [32,33]. Even without DAO, gEUD introduces a complicated non-quadratic structure into the objective function. Therefore, a stochastic optimization is indispensable for such type of objectives to avoid local minima as solutions.

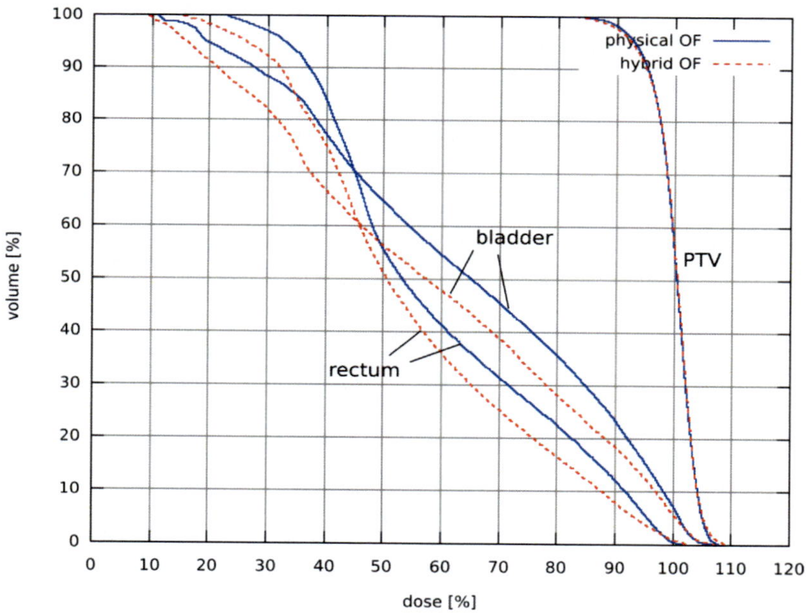

Figure 5. Potential advantage of biological over physical optimization demonstrated for a prostate cancer case. Normal tissue may be better spared applying a hybrid objective function (OF) for optimization with a gEUD-based OF for the tumor (PTV) and a physical OF for the organs at risk [33].

Biological objective functions tend to spare normal tissue in a high extent, but simultaneously allow slight underdosage in tumor subvolumes.

This is due to the fact that small underdosed tumor volumes do not remarkably contribute to the scoring value in case of gEUD-based objective functions, but it diminishes TCP dramatically. Therefore, hybrid objectives should be preferable, which means that the tumor part is optimized according to physical objectives whereas the normal tissue part according to biological objectives [32,33].

$$f = \sum_{oar} (gEUD^{oar} - gEUD_0^{oar})^2 + f_{phys}^{ptv} \quad \text{eq. (9)}$$

By this technique, improvements in normal tissue sparing over pure physical optimization are possible while simultaneously target dose coverage can be maintained high [33], as shown for a prostate cancer case in Figure 5.

2.4. Geometrical Uncertainty: Improvement by IGART

Inter- and Intrafractional Uncertainties

Treatment planning assumes per se the static conditions at the time of the CT-scan. But most of all, tumor irradiations consist in a fractionated treatment scheme, e.g. typically 30-40 daily treatments with a dose per fraction of about 1.8-2.0 Gy. Therefore, the time course of daily patient positioning as well as volume changes and target movements during irradiation have to be introduced in an overall quality management system of the radiotherapy process. The first two parts can be sub-summarized to yield an inter-fractional positioning error and the latter to yield an intra-fractional movement error.

Stereotactic Treatment

A very powerful treatment method to minimize inter- and intra-fractional geometrical errors are stereotactic treatments. Stereotactic treatment was introduced for so-called hypofractionated irradiations of brain-tumors or h&n-tumors in only 1 to 3 fractions in a region of the body, which exhibits almost no internal tissue movement. The remaining inter-fractional positioning error can reduced dramatically by a stereotactic frame, which is fixed invasively by 4 screws on the patient's bony structure. The frame defines a highly precise coordinate system fixed rigidly to the patients skull. As it remains fixed during the whole treatment chain from diagnostics to treatment, one is able to reduce positioning errors in the submilimeter range [34]. A similar method was extended to extra-kranial areas of the body, e.g. in stereotactic treatments of lung or liver tumors. But as precision in the body stem cannot be maintained at the high level of brain tumor stereotaxis, another method termed image-guided-radiotherapy (IGRT) or frameless stereotaxis was developed.

Tumor Tracking in Image-Guided Radiotherapy

Among the most challenging tumor sites in the human body are tumors of the thorax, and the pelvic region, e.g. prostate cancer. Tumors within the thorax present the highest amplitudes of local movement on breathing while tumors in the pelvic region are, within most of the treatment time, stable in position, but this may change unexpectedly by peristaltic movement of the rectum. Nowadays, linear accelerators are furnished with a complete diagnostic equipment for fast acquisition of so-called cone-beam CTs with kilovoltage X-ray tubes and 2D-detectors. By such a CT scan, fast checking of the target in actual treatment setup is possible prior to each treatment fraction. The adaptation of the linear accelerator to correct for the detected positioning errors will be described below.

Another detection method makes use of electronic portal imaging devices (EPID). They consist of amorphous silicon array detectors with high spatial resolution in combination with kilovoltage X-ray tubes. By the use of two such equipments, a stereotactic view of the tumor-containing environment of the patient is possible. For both X-ray beam directions so-called digitally reconstructed radiograms (DRR) have to be precalculated during treatment planning. DRR is a simulated planar X-ray view along the choosen directions, which can be calculated by the treatment planning system from the 3D-data set of the patient's CT investigation [35]. Tumor tracking is then possible by deriving quantitatively deviation parameters from both EPID-images compared to the corresponding DRRs. The evaluation of the comparisons of the measured images with the calculated ones is done by image-registration algorithms. In most cases, rigid registration methods, where the target is regarded as incompressible, the mutual information metric algorithm (Rosenman in [36]) is applied. By that, a fast detection of the deviations of the target along translational and rotational degrees of freedom is possible. Non-rigid registration algorithms are currently under development, which take into account the compressibility of the tumor target.

High precision linear accelerators are supplied with so-called hexaPod-couch mounts. This is a table top mounted on top of the treatment couch which offers patient positioning by three translational and three rotational degrees of freedom with very high precision.

After application of registration algorithms to on-site X-ray investigations and images from treatment planning (either CTs or DRRs), deviation parameters are available for automatical adjustment of translational and rotational parameters by the hexaPod-couch.

Thus, IGRT (or also termed *image-guided adaptive-radiotherapy* (IGART) is able to correct for interfractional target mispositioning for static conditions during one treatment fraction, which lasts a few minutes. There are tumor sites with pronounced geometrical dynamics even during one fraction, caused e.g. by breathing, heart beat or peristaltic movement of the intestine. Periodic movements e.g. in thorax sites can be detected by optical signals from markers, fixed to the skin surface. But this skin movement differs from target movement.

Therefore, in a study performed prior to treatment, a movement function describing the correlation between inner and outer movements has to be derived. This is possible since dynamic CT-scanners are available, which allow the quantitative detection of target movement. Now, during irradiation, the skin movement function is on-line detected and translated into target movement by multiplication with the movement function to serve as a time-dependent signal for tumor tracking. Tracking may either be performed by applying the hexaPod couch mount dynamically, inducing a moving patient or by dynamic movement of the MLC-leafs, controlled by the tracking signal in both cases.

Another way to increase precision is to trigger the linear accelerator output according to thresholds on the target movement function. Setting the threshold narrow to the deep inspiration or exhalation values and allowing beam-on only above or below that threshold, respectively, the tumor is only treated in the correct position.

As the price for this precision is an enlargement of treatment time, one has to find a compromise between both factors. The most challenging source of imprecision is peristaltic movement in the pelvis during irradiation. Most promising are inverse planning algorithms which apply probabilistic optimization. From a whole series of CT-scans over a distinct time period of several days, the distribution of the probability that the tumor resides within a distinct voxel is derived. This information enters into the objective function to modulate the importance weight of the voxels of the patient volume with this probability distribution. Such the probability that the tumor receives the prescribed dose is maximized [36].

2.4. A Paradigma Change - Handling Heterogeneous Tumor Biology

Molecular Imaging

Along with new imaging modalities like single photon emission tomography (SPECT) or positron emission tomography (PET) as well as functional magnetic resonance tomography (fMRT) or magnetic resonance spectroscopy (MRS), a deeper inside in biological tissue properties can be gained. These modalities are very interesting for radiotherapy, as they are able to detect distinct tumor properties with a high spatial resolution. Along with the development of new tracers for SPECT and PET properties like radioresistence, hypoxia (also leading to radioresistance) and aggressiveness or cell proliferation rate (as detected by amino acid turnover) or simply cell density, can now be detected non-invasively by 3D-imaging. One remaining problem to be resolved for each tracer is the exact knowledge of a local tumor property, when a distinct biosignal is detected there. There are several tracers currently interpreted as surrogate for tumor properties like hypoxia (FAZA, FMISO) or proliferation (IMT, FMT, FET, FCH) [38]. Other molecular imaging methods like fMRT or MRS come up with similar 3D-biosignal information like hypoxia or proliferation and cell density.

Multimodality Target Definition and Biological–Eye-View

Ling et al. [30] suggested in an early publication to include all of the multimodal image information into a new paradigma of the tumor which they termed *biological-eye-view*. He suggested to apply a locally varying dose to different target subvolumes according to their biological properties. This may be performed by manual methods such as applying a local boost-irradiation to sub-volumes. But the molecular imaging methods allow the development of much more elegant techniques by a modified IMRT technique.

Optimizing Biological Effect Rather Than Dose - BART

In principle, a locally varying dose prescription can be derived from molecular imaging information, if one is able to correlate the local biosignal to a dose prescription, which is necessary for a complete tumor destruction. This technique may be termed biologically adaptive radiotherapy (BART). The heterogeneously prescribed dose can easily be used in the inverse planning process to achieve the proper local dose. As a first attempt to resolve the problem, a linear correlation between prescribed dose and biosignal was suggested [39,40,41].

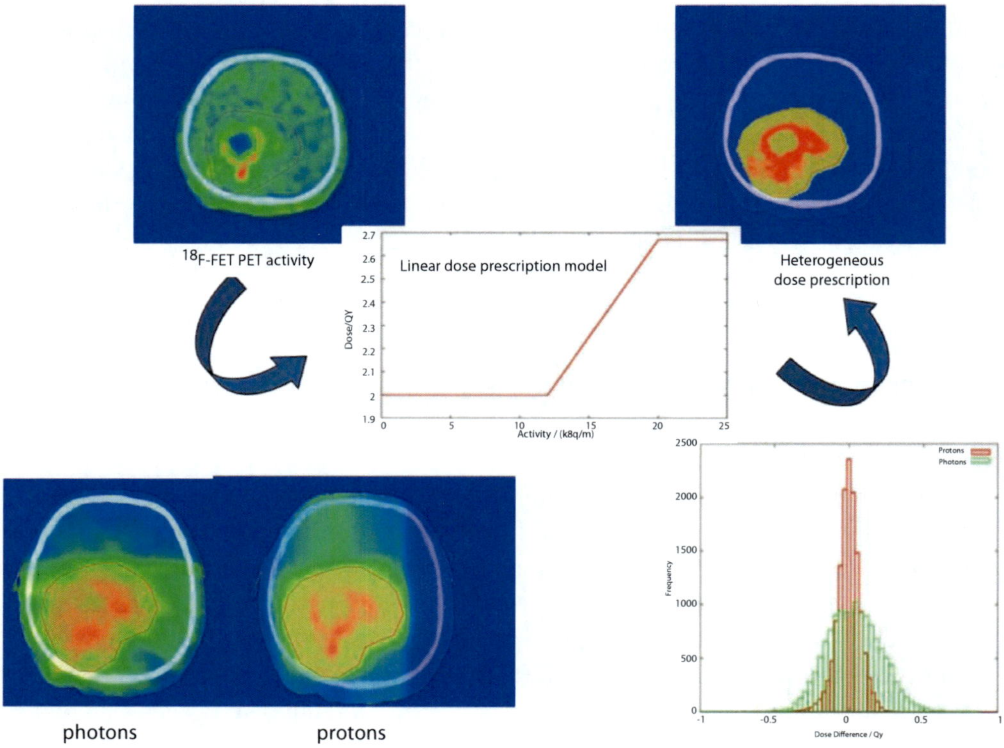

Figure 6. Biologically adaptive radiotherapy (BART) of a brain tumor, based on ^{18}F-FET PET. Top: The images containing the PET biosignal are translated into a heterogeneous dose prescription by means of a simple linear model. Bottom: Optimization is accomplished by DMCO for photons and IKO for protons. The dose difference histogram (photon and proton dose, respectively, minus dose prescription) demonstrates clearly the advantage of protons [41,50].

The application of this model is shown in Figure 6 for a brain tumor case, where after surgery, a heterogeneous pattern of biosignal was detected in the PET investigation with a ^{18}F-FET-tracer, such indicating tumor activity within the remaining tissue. Such a model allows a remarkable local dose escalation inside the tumor while simultaneously, the dose to organs-at-risk remains almost unchanged [40].

In a first approach, a linear model was set up by introducing values for the minimal dose for tumor destruction, and a hypothetical maximal escalation dose, derived from clinically reliable knowledge. This linear model allows the construction of a heterogeneous dose prescription for the optimization to be performed by IKO. A histogram of the difference between calculated and prescribed dose reveals a rather good agreement.

Obviously, biological effects seldom obey simple linear relationships with dose. Therefore, another, more realistic biological approach tries to achieve a constant tumor control probability, irrespective of its heterogeneous properties. Xing and Yang [52] proposed such a model for macroscopic sub-volumes of the tumor by applying the LQ-model which was amended by terms describing the dependence of TCP from cell density ρ and cell proliferation rate ρ additional to the LQ-model parameters α and α/β. The model can be generalized to derive a voxel-wise dose-prescription under the constraint of a homogeneous TCP [41]. Investigations of how the model parameters described above influence TCP show, that for brain tumors, the most influence is exerted by the LQ-parameters α and α/β.

Postulating that α and α/β depend linearly on the biosignal, a dose prescription for each voxel number i can be derived by

$$D_0^T(i) = \frac{\alpha'_{ref}}{\alpha'_i} D_{ref} - \frac{1}{\alpha'_i}(\gamma_{ref} - \gamma_i)\Delta T - \frac{1}{\alpha'_i}\ln\left(\frac{\alpha'_{ref}\rho_{ref}}{\alpha'_i \rho_i}\right) \qquad \text{eq. (10)}$$

The voxel-wise dose prescription depends non-linearly on the biosignal and it was derived under the constraint of a constant TCP. Figure 7 shows a prostate case optimized under the constraint of constant TCP, revealing the strong influence of small underdosed subvolumes within the target volume on TCP. Achieving high tumor control even under such conditions forces to escalate the maximal regional dose to extreme values.

The TCP-model can easily be adapted to other tumors by taking into account the different tumor-specific LQ-model parameters and investigation of the sensitivity of the different terms on TCP.

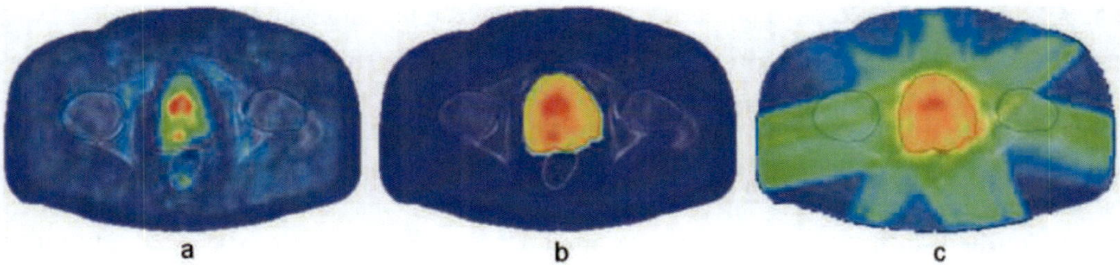

Figure 7. TCP-based dose-painting-by numbers of a prostate case. a) Distribution of ^{18}F-Choline PET-activity. b) Heterogeneous dose prescription, derived by the TCP model delivers a non-linear dependence of γ and ρ from the activity. c) Resulting dose distribution after optimization with DMCO[42].

A BART-study with ^{18}F-Cholin PET on a prostate cancer showed, that the modulation follows the dose prescription only up to a distinct amount, which is defined by the modulation ability of the treatment modality [42]. This ability is restricted depending on the minimal segment shape, which can be applied compared to the structure sizes to be modulated. Therefore, prostate cancer has to be treated with higher resolution MLCs.

BART is in its early stage of research and it strongly depends on new findings and comprehension of the biological nature of the biosignals. Once the comprehension is established, improved biological models for biosignals have to be developed. This, as a vision, a tremendous improvement of treatment outcome may be postulated.

2.5. New Beam Qualities - From Photons to Heavy Charged Particles

High energy photons in the MV range are mostly applied in radiotherapy, because a large penetration depth is achievable and skin surface can be spared by the so-called build-up effect in the depth dose curve. In contradiction, heavy charged particles present a completely different behavior as they show the so-called Braggpeak. The depth dose curve of protons or heavy ions is termed *inverse* because the dose near the entrance into the tissue is low but increasing with depth until it reaches the range of the particles. Beyond that range, a rapid fall-off of the dose is observed. The reason for that behavior is the multiple Coulomb interaction of the charged particle with atomic electrons, which leads to a very small energy loss per each event. After as many as a few thousand interactions, the particle comes to rest and with that, the interaction time increases. This leads to a remarkable increase in dose deposition at the Braggpeak [43]. Distal the Braggpeak no dose deposition occurs. Beside this interesting effect, as it concerns the application to radiotherapy, another positive effect is the lateral beam precision. As the mass of particles such as protons up to heavier ions such as carbon is very high, the lateral deflection is low, increasing with the atomic number [44,45].

Radiotherapy with protons and heavy ions started several decades ago at nuclear research facilities, whereas the first clinically dedicated facilities started in the early 1990s with protons. About ten years later, clinical carbon ion projects started. Most of the treatment facilities applied passive scattering and energy modulation techniques to produce a Braggpeak, which spreads out over tumor volume. This method has disadvantages compared to spot scanning methods introduced first by the PSI [46]. They direct the Braggpeaks of narrow proton pencil beams to predefined positions within the target and cover the target with dose in a step-and-shoot procedure. The intensity of each spot is modulated in order to achieve a homogeneous dose within the target. Lomax [47] showed that intensity modulated proton therapy (IMPT) based on inverse planning can improve dose distributions remarkably. The planning system made use of a simple pencil beam dose algorithm for protons. In proton dose calculation and even more in applications to heavier ions, different sources of imprecision and uncertainties arise, which may deteriorate the treatment considerably. One difficulty is the exact determination of the Braggpeak position in heterogeneous tissues. Deviations from the precalculated particle range may lead to underdosage of tumor tissue or overdosage in organs at risk.

Additionally, because of the stochastic nature of energy deposition within several thousands of interactions proximal the Braggpeak, particle paths may "diffuse" along different pathways in the tissue. This leads to different histories with respect to tissue heterogeneities, such yielding a smearing-out or Braggpeak-degradation, as it is termed, instead of sharp single Braggpeaks.

Another drawback of PB-algorithms is that nuclear reactions are only accounted in a simplistic way. MC dose algorithms take into account all of these issues with a very high degree of precision, such as the proton-tissue interaction in heterogeneous tissues or nuclear interactions. Diminishing dose errors on the other hand, reduces the convergence error in inverse planning. A very precise and simultaneously fast MC code for protons termed 'VMCpro' was developed by Fippel et al. [48].

This code was applied to derive the inverse kernel matrix in an IKO concept for protons by Moravek et al. [49], which can be used for inverse kernel optimization. As uncertainty is a crucial point in IMPT, further attempts are integrated in the approach to minimize those. Beside reduction of systematic dose errors and the subsequently induced convergence errors, uncertainty can be reduced by strategies, that avoid crucial beam directions. To achieve that goal, a beams angle optimization part and additionally, a spot planning part have been introduced. Optimization of the beam angles is aspired by means of an objective function, which includes a term to put penalty on particle paths through large volumes of organs-at-risk.

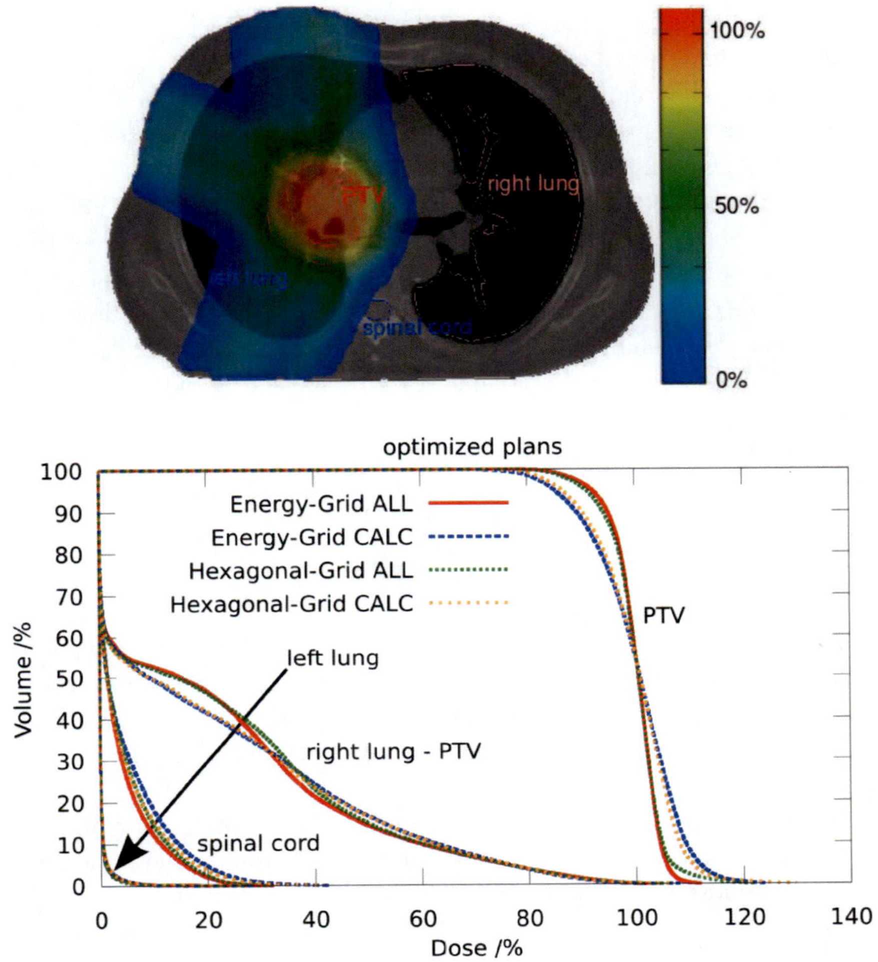

Figure 8. Intensity-modulated proton beam plan for a lung cancer case with 3 beams, optimized with IKO-P. Top: Transversal CT-slice with the dose distribution. Bottom: Integral dose-volume-histogram (DVH) for the planning target volume (PTV) and organs at risk. Different options for the spot planning are included [49].

Secondly, two other terms increase score value when the path encounters regions of high tissue heterogeneities, particularly if the variance in tissue density parallel to the path and additionally transversal to the path is high. This diminishes problems with range uncertainty or Braggpeak degradation. Finally, penalty is put on beam overlap to avoid similar beam directions.

Another source of uncertainty may occur on spot planning. Spot planning means the predefinition of spots which shall be reached by the Braggpeaks of the pencil beams. To assure this precisely, a grid to put spots on is defined over the target and pencil beams are selected by a ray trace procedure. There are several degrees of freedom to bring dose to the spot as to be i) from all directions or ii) from optimized directions according to the beams angle optimization objective function. By the latter option, directions are chosen that try to avoid areas of pronounced heterogeneities and organs at risk. Another issue, introduced into IKO-P is a free choice among different grid types that exhibit different amounts of space filling.

Investigations revealed that grids associated with a higher space filling factors present a smaller variance in the distribution of the spot-weights after optimization. This may be an advantage for tumor sites in moving environments, as similar weights of neighbored spots are less sensitive to tissue movements over different pronounced weighted spots.

A lung cancer case optimized with IKO-P is shown in Figure 8 with options for an optimized beam direction (CALC) or a contribution of all beams to every spot (ALL) respectively. Additionally two different spot-grids are demonstrated. A comparison of the evaluations of the optimization with protons (Figure 8) compared to photons (Figure 1) demonstrates the superiority of IMPT over IMRT especially for the normal tissue.

Protons have a higher ability to conform dose to the target and simultaneously to bring dose to normal tissue down to roughly a factor of two compared to photons. Additionally, the modulation capability is increased, which could be shown by Rickhey et al. [50] to be advantageous in BART.

Conclusion

Optimization in high precision radiotherapy needs first and foremost the highest accuracy in dose calculation to minimize systematic and dose errors. The code of choice is MC simulation, as its accuracy is only a question of the number of simulated histories. Furthermore, the coverage of heterogeneous tissue structures is, by far, most precisely handled by MC simulation.

Inverse planning is another technique to achieve highly conformal dose distributions that meets the clinically defined constraints. Optimization algorithms can either be standard deterministic, when dose-based objective functions are applied. Promising are biologically objective functions, but which necessitate stochastic algorithms like e.g. simulated annealing. Hence, inverse MC optimization is potentially in favor over all other conventional algorithms. High precision treatment planning inforces to maintain accuracy not only during planning, but also on irradiation of each single fraction as well.

Therefore, image-guidance is a prerequisite in the application of these techniques. More and more strategies to reduce uncertainties have additionally been subjects of robust planning and dose administration. Regarding the historical development of radiotherapy, its progress from 3DCRT and IMRT with photons to IMRT with protons and heavy ions can be imagined in a dose-volume histogram, such demonstrating the success on our way to the infinitron [13] (Figure 1). An overview of innovative technologies in radiotherapy is given by S. Webb and P. Evans [51].

But finally, one has to be aware that escalation of dose to thetumor and simultaneously reduction of dose to the normal tissues is not primarily a physical question. It is a biological problem, as to how tumor tissues in their histological range and the characteristics of the different normal tissues respond to radiation dose. Hence, coming back to the roots of radiotherapy, the future will be BART. But there is some way in front of us to penetrate much deeper into knowledge about the question, how a biosignal has to be interpreted. An answer to questions like this must come from molecular biology and physiology.

References

[1] Smith, A.R. (editor). Radiation therapy physics, Springer, Heidelberg, ISBN 3.540-55430-0 (1995).

[2] Khan, F.M. The physics of radiation therapy (2. ed.). Wiliwam and Wilkins, ISBN 0-683-04502-4 (1994).

[3] Ahnesjö,A., Aspradakis, M.M. *Phys. Med. Biol.* 44 R99 (1999).

[4] Bielajew,A.F., Rogers, D.W. *Med. Phys.* 19(2), 303 (1992).

[5] NRC National Research Council Canada. Publications, http://www.irs.inms.nrc.ca/papers/irs_www/irs_www.html .

[6] Fippel, M. *Med. Phys.* 26(8),1466 (1999).

[7] Kawrakow, I. *Med. Phys.* 27(3), 485 (2000).

[8] Rogers, W.D.O., Faddegon, B.A., Ding,G.X., et al. *Med. Phys.* 22, 5031995).

[9] Fippel, M., Haryanto, F., Dohm, O., et al. *Med. Phys.* 30(3), 503 (2003).

[10] Jeraj, R., Keall, P.J., Siebers, J.V. *Phys. Med. Biol.* 47(3), 391 (2002).

[11] Chetty, I.J., Curran, B., Cygler, J.E., et al. *Med. Phys.* 34(12), 4818 (2007).

[12] Bortfeld, T. *Semin. Radiat. Oncol.* 9(1), 20 (1999).

[13] Schulz,R.J., Kagan, A.R. *Med. Phys.* 30(2);276 (2003).

[14] Bortfeld, T. *Phys. Med. Biol.* 51(13), R363 (2006).

[15] Alber, M., Nuesslin, F. *Phys. Med. Biol. 45(*5), N49 (2000).

[16] Rickhey, M., Moravek, Z., Bogner L. *Z. Med. Phys. 18*(3), 163 (2008).

[17] Bogner, L., Hartmann, M., Rickhey, M., Moravek, *Med. Phys.* 33(12), 4749 (2006).

[18] Laub, W., Alber, M., Birkner, M., et al. *Phys. Med. Biol* 45(7), 1741 (2000).

[19] Bogner, L., Scherer, J., Herbst, M. *Physica Medica* XV(3), 111 (1999).

[20] Hartmann, M., Bogner, L., Fippel, M., et al. *Z. Med. Phys.* 12, 97 (2002).

[21] Gustafsson, A., Lind, B.K., Brahme, A. *Med. Phys.* 21(3), 343 (1994).

[22] Bergmann, A.M., Bush, K., Milette,M.P., et al. *Med. Phys.* 33(10), 3666 (2006).

[23] Shepard, D.M., Earl, M.A., Li, X.A., et al. *Med. Phys.* 29(6), 1007 (2002).

[24] Webb, S. *Phys. Med. Biol.* 34(10), 1349 (1989).

[25] Schneider, J.J., Kirkpatrick, S. (editors). Stochastic optimization, Springer, Berlin, Heidelberg, ISBN-10 3-540-34559-0 (2006).

[26] Bogner, L. ,Alt, M., Dirscherl,T., et. al. *Phys. Med. Biol.* 54(13), 4051 (2009).

[27] Bush, K., Popescu, I.A., Zavgorodni, S. *Phys. Med. Biol.* 53, N337 (2008).

[28] Jeraj, R.,Keall, P. *Phys. Med. Biol.* 44(8), 1885 (1999).

[29] Niemierko, A. *Med. Phys.* 24 103 (1997).

[30] Ling, C.C., Humm, J., Larson, S., et al. *Int. J. Radiat. Oncol. Biol. Phys.* 47(3), 551 (2000).

[31] Niemierko, A. *Med. Phys.* 26, 1100 (1999).

[32] Hartmann, M., Bogner, L. *Med. Phys.* 35(5), 2041 (2008).

[33] Dirscherl, T., Alvarez-Moret, J., Bogner, L. Z. *Med. Phys.,* 21(3), 228 (2010).

[34] Treuer, H., Hoevels, M., Luyken, K., et al. *Acta Oncol.* 42(2), 92 (2003).

[35] Jin, J.Y., Yin, F.F., Tenn, S.E., et al. *Med. Dosim.* 33(2);124 (2008).

[36] Xing, L., Ling, C., Mitchell, J.B. (editors.) *Semin. Radiat. Oncol.* 11(1) (2001).

[37] Witte, M.G., van der Geer, J., Schneider, C., et al. *Med. Phys.* 34(9), 3544 (2007).

[38] Grosu, A.L., Piert, M., Weber, W.A. *Strahlenther Onkol* 181(8), 483 (2005).

[39] Thorwarth, D., Alber, M. *Z. Med. Phys.* 18(1), 43 (2008).
[40] Rickhey, M., Koelbl, O., Eilles, C., Bogner, L. *Strahlenther. Onkol.* 184(10), 536 (2008).
[41] Rickhey, M. Dissertation, University of Regensburg, http://www.opus-bayern.de/uni-regensburg/volltexte/2009/1117/ (2009).
[42] Dirscherl., T., Rickhey, M., Bogner, L. *Z. Med. Phys.*, ahead to print under DOI 10.1016/j.zemedi.2011.09.006 (2011).
[43] Moravek, Z., Bogner, L. *Z. Med. Phys.* 19(3), 174 (2009).
[44] Amaldi, H., *Z. Med. Phys.* 14(1), 7 (2004).
[45] Goitein, M. Radiother. Oncol. 95(1), 23 (2010).
[46] Pedroni, E., Scheib, S., Böhringer, T., et al. *Phys. Med. Biol.* 50(3), 541 (2005).
[47] Lomax, A.J., Pedroni, E., Rutz, H., et al. *Z. Med. Phys.* 14(3), 147 (2004).
[48] Fippel, M., Soukup, M. *Med. Phys.* 31(8), 2263 (2004).
[49] Moravek, Z., Rickhey, M., Hartmann, M., Bogner, L. *Phys. Med. Biol.* 54(15), 4803 (2009).
[50] Rickhey, M., Moravek, Z., Eilles, C., Kölbl, O., Bogner, L. *Strahlenther. Onkol.* 186(6), 320 (2010).
[51] Webb, S., Evans, P. *Semin. Radiat. Oncol.* 16(4), 193 (2006).
[52] Yang, Y., Xing, L. *Med. Phys.* 32, 1473 (2005).
[53] Alt, M. Diploma thesis, University of Applied Sciences, Regensburg, http://epub.uni-regensburg.de/10742/1/DiplArb_Alt_Final.pdf (2008).

In: Medical Physics
Editor: Marcin Balcerzyk
ISBN: 978-1-62257-590-9

Chapter VIII

Computer-Aided Detection and Diagnosis for Interstitial Lung Disease in Medical Imaging

***Jiahui Wang*[1,2] *and Qiang Li*[1,*]**
[1]Department of Radiology, Duke University, Durham, NC, US
[2]Department of Psychiatry, The University of North Carolina, Chapel Hill, NC, US

Abstract

Interstitial lung disease (ILD) is a common pulmonary disorder that affects the tissue and space around the air sacs of the lungs. Most ILD patients are diagnosed with blood testing and pulmonary function testing. Because chest radiography and computed tomography (CT) can provide a non-invasive assessment of the suspected regions of lesions and have the ability to visualize the entire lungs in great detail, they have been widely used for diagnosis of ILD. In particular, as CT technologies have advanced, CT has become the most common imaging modality for clinical practice and research of ILD.

Although high-resolution medical images produced by modern medical imaging equipments are valuable resources for the diagnosis and treatment assessment, subjective reading and analysis of the large amount of the image data are tedious, time-consuming, and subject to human reader variability. Furthermore, it is a very difficult task for radiologists to correctly diagnose patients with ILD in medical images, because the contrast of lesions is often low and the disease patterns are very complex. Therefore, computer-aided detection and diagnosis (CAD) schemes have been developed for assisting radiologists in the detection and diagnosis of ILD.

This article reviews the current status on CAD schemes of ILD in chest radiography and CT and addresses some key questions in current CAD schemes of ILD, including segmentation of lungs, feature analysis, classification and quantification of ILD, as well as the evaluation of performance levels. Our review work on the CAD schemes of ILD

* Correspondence to: li.qiang@duke.edu

will help researchers to understand the current status and research direction in this field. We believe that this article would be a valuable resource for development and evaluation of computerized schemes for ILD in the future.

1. Introduction

Interstitial lung disease (ILD) is a common pulmonary disorder that affects the tissue and space around the air sacs of the lungs. Most ILD patients are diagnosed with blood testing and pulmonary function testing. Because chest radiography and computed tomography (CT) can provide a non-invasive assessment of the suspected regions of lesions and have the ability to visualize the entire lungs in great detail, they have been widely used for the diagnosis of ILD. In particular, as CT technologies have advanced, CT has become the most common imaging modality used in clinical practice and research of ILD. However, subjective reading and analysis of the large amount of the CT image data are tedious, time-consuming, and subject to human reader variability.

Furthermore, it is a very difficult task for radiologists to correctly diagnose patients with ILD in medical images, because the contrast of lesions is often low and the disease patterns are very complex.

Therefore, computer-aided diagnostic (CAD) schemes have been developed for assisting radiologists in the detection and diagnosis of ILD. This article reviews the current status on CAD schemes for ILD in chest radiography and CT and addresses some key issues in current CAD schemes, including the segmentation of lungs, feature analysis, classification and quantification of ILD, as well as the evaluation of performance levels. This review article will help researchers understand the current status and research direction in the CAD schemes for ILD, and thus would be a valuable resource for development and evaluation of computerized schemes for ILD in the future.

The publications reviewed in this article were obtained from online research database search engine PubMed from the 1980s through 2011. The key words used to search the publications were (“computer-aided diagnosis” or “computer-aided detection”) and (“interstitial lung disease” or “diffuse lung disease”).

2. Overview of Interstitial Lung Disease

Interstitial lung disease, also called diffuse parenchymal lung disease or infiltrative lung disease, is a common thoracic disorder that primarily affects the lung parenchyma and space around the air sacs of the lungs, and varies widely in etiology, clinic radiologic presentation, histopathologic features, and clinical course [1]-[4]. It includes more than 200 chronic lung disorders [5], which can cause progressive scarring of lung tissue and result in impaired gas exchange, respiratory insufficiency, and ultimately death due to respiratory failure. One study reported that 80.9 per 100,000 men and 67.2 per 100,000 women suffer from ILD in the United States with 31.5 new cases diagnosed per 100,000 men per year and 26.1 new cases diagnosed per 100,000 women per year [6].

2.1. Disease Patterns

Interstitial lung disease usually presents as texture-like intensity changes in chest radiography or CT images [7] due to an abnormal accumulation of fluid, cells, or other tissue elements within the small air spaces or interstitium. The most common ILD patterns are ground glass opacity, nodular pattern, reticular pattern, honeycombing pattern, emphysema, and consolidation.

Ground-Glass Opacity

The term "ground-glass" opacity (GGO) indicates a common finding on chest radiography or CT, characterized by areas of increased attenuation of the lung parenchyma with poor visibility of surrounding structures [8]. Figure 1A shows an portion of a slice of multi-detector CT images of a patient with GGO. GGO pattern can be observed in a large number of ILDs. By taking into account its distribution, the associated findings, and the clinical parameters, it is possible to make a correct diagnosis for most ILDs with GGO pattern [9].

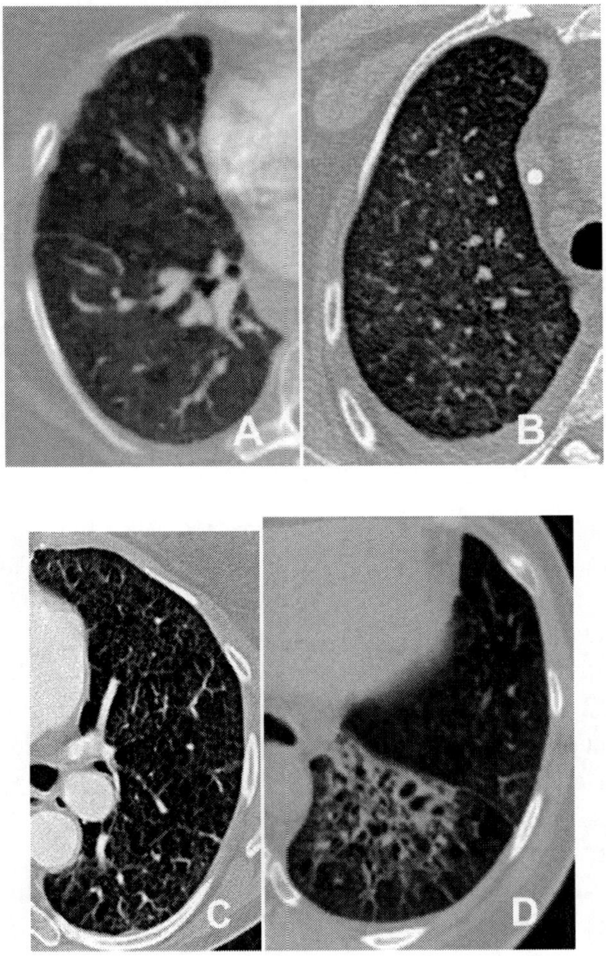

Figure 1. (Continued).

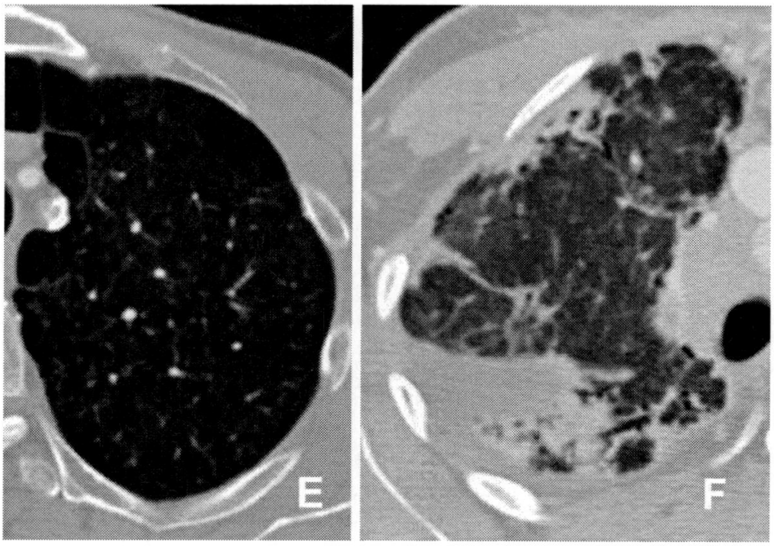

Figure 1. Illustrations of multi-detector row CT images for different patterns of interstitial lung diseases, including A. ground glass opacity, B. nodular pattern, C. reticular pattern, D. honeycombing pattern, E. emphysema, and F. consolidation.

Nodular Pattern

Nodular pattern appears as multiple small circular opacities (usually smaller than 1 cm in diameter) [10] and shows a blocky distribution on chest radiography or CT [Figure 1B]. Nodules can be classified as "perilymphatic nodules", "centrilobular nodules," and "random nodules" based on their distribution and locations [9].

Reticular and Honeycombing Patterns

The term reticular pattern is purely descriptive and there are several variations in appearance for this basic pattern, ranging from thickening of the interlobular septa [Figure 1 C] to honeycombing lung destruction [Figure 1 D] [11].

Emphysema

Emphysema is a long-term, progressive disease of the lungs that primarily causes shortness of breath due to over-inflation of the alveoli (air sacs in the lung). Smoking is the major cause of emphysema. In people with emphysema, the lung tissue involved in exchange of gases (oxygen and carbon dioxide) is impaired or destroyed. Emphysema turns the spherical air sacs into large and irregular pockets with gaping holes (low intensity holes in chest radiography and CT) in their inner walls and reduces the number of air sacs, as shown in Figure 1 E. Although emphysema can be regarded as an independent lung disease, it is often classified as a pattern of ILD.

Consolidation Pattern

Consolidation is defined as an area of highly increased attenuation with obscuration of the adjacent bronchial walls and vessels [Figure 1 F].

2.2. Imaging Modalities Used for Interstitial Lung Disease

The first radiological diagnostic approach to ILD is based on chest X-ray radiography. Since then, the standard chest radiography remains unsurpassed in terms of cost, radiation dose, availability, and ease of performance. However, the chest radiography has several limitations, including the superimposition of lesions and normal anatomic structures and the relatively low resolution in contrast. Computed tomography (CT) overcomes these problems and allows for a better assessment of the type, distribution, and severity of parenchymal abnormalities than does the chest radiography. The accuracy of high resolution CT (HRCT) for detecting disease and distinguishing among different diseases has been well documented, and its advantages over chest radiography [12]-[14] and conventional CT have been elucidated [12],[15]. HRCT provides an optimal assessment of parenchymal details that closely reflect the microscopic pathologic findings [16],[17]. Presently, new advances in multi-detector CT (MDCT) technology may allow routine sub-millimeter scanning [12], total lung volumetric high-resolution CT [18], multi-planar reconstructions [19], dynamic CT throughout the respiratory cycle, and reproducible spirometric gating of data acquisitions at specific points in the respiratory cycle [20].

3. CAD Schemes for Interstitial Lung Disease

Diagnosis of ILD is one of the most difficult tasks for radiologists [21]. Therefore, CAD systems have been developed for assisting radiologists in the detection and diagnosis of ILD. Radiologists can use the output of CAD schemes as a "second opinion" to make differential diagnosis of various diseases using medical images [22]-[24]. The studies on CAD for ILD begun in the 1980s at the Kurt Rossmann Laboratory for Radiologic Image Research in the Department of Radiology at the University of Chicago [25]. Early CAD schemes for ILD were developed on chest radiography. As the progress of CT technology, CT has been widely used in the development of CAD schemes for ILD. In particular, most of the studies published after 2000 were conducted on CT. Figure 2 shows the number of publications regarding CAD schemes for ILD using chest radiography and CT in different periods of time.

CAD schemes for ILD can be divided into two categories, namely, computer-aided detection (CADe) and computer-aided diagnosis (CADx). Although CADe and CADx for ILD are implemented in a similar manner, the goal of CADe is to identify and mark suspicious areas in images, whereas the goal of CADx is to classify the abnormal area into different ILD diseases.

Thus, the major difference between CADe and CADx exists in their outputs. The output of a CADe scheme is usually represented by many symbols/marks placed on the images to indicate where the abnormal lung parenchyma is, and the output of a CADx usually indicates the likelihood values of being specific ILD diseases [26]. For the CAD studies reviewed in this article, there are 6 CADe and 11 CADx schemes developed in chest radiography, respectively, and 9 CADe and 13 CADx schemes in CT, respectively.

From technical point of view, CADe and CADx for ILD have similar structures. In order to simplify our description, we will not describe CADe and CADx schemes separately in different sections.

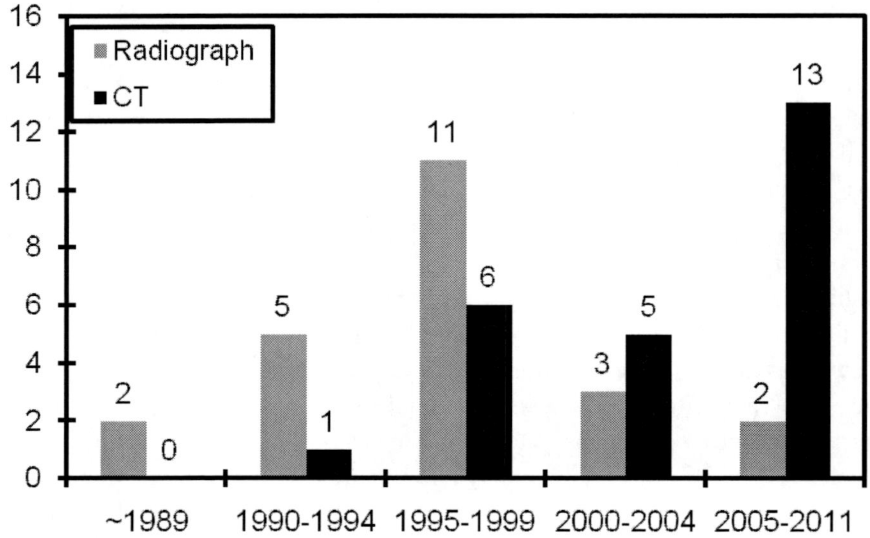

Figure 2. The numbers of CAD schemes for ILD developed in chest radiography and CT during different periods of time.

CAD schemes are fundamentally based on image processing, pattern recognition, and artificial intelligence techniques. The overall pipeline of a CAD scheme for ILD is generally composed of five steps: (1) pre-processing for reducing artifacts and image noise; (2) segmentation of interested structures in medical images, such as the lung; (3) determination of region of interest (ROI); (4) image-feature extraction and analysis in ROIs; and (5) classification of ROIs based on the image features for identification of normal and abnormal ROIs. Although the pre-processing step is an important component, it may vary widely for different CAD schemes and even may not be needed. Therefore, we will focus on the other four major steps. We will also review the observer performance studies for demonstrating the usefulness of CAD schemes in the detection and diagnosis of ILD.

3.1. Image Segmentation

Segmentation of Ribs

Image segmentation is usually an important first step to many applications for medical image analysis. For the CAD schemes for ILD, the segmentation of ribs, pulmonary vessels, airways, and lung regions are often required. Because ribs may cause a large number of false positives in the detection of ILD, correct identification of ribs could considerably improve the accuracy of CAD schemes.

The thresholding technique was first used to segment ribs in chest radiography [27]. However, simple thresholding might not work well when the lungs contain high intensity ILD patterns such as GGO. Therefore, Ishida et al. [28] proposed a feature analysis-based technique to remove low-contrast ribs. Rib edges were first detected from pixels whose edge gradients exceeded a predetermined threshold. Because the low-contrast rib edges might not be identified by this simple strategy, a rule-based classification was performed to identify the low-contrast rib edges by use of the line components and area components. In Rudrapatna et al.'s study [29], the ribs candidates were identified from CT images by a simple thresholding technique (fixed threshold = 250 HU). Rib candidates containing a black hole were classified as true ribs, since ribs enclose soft tissue of bone marrow, which appears as a black region in CT images.

Segmentation of Pulmonary Vessels and Airways

Pulmonary vessels and airways are also major sources of false positives in the detection of ILD. Chabat et al. [30] employed mathematic morphology to develop a structure-filtering operator for identifying and removing major pulmonary vessels from chest CT. Zavaletta et al. [31] proposed an airway segmentation method, which used region-growing algorithm with multiple thresholds, to optimally extract airways. They also segmented pulmonary vessels by filtering the CT images with a 3D line-enhancement filter, which is based on the examination of the eigenvalues of Hessian matrix [32]. Marten et al. [33],[34] also proposed an airway segmentation method using region-growing technique. A graph presentation of the segmented airways was constructed by use of the information about the direction, diameter and length of each branch of the airways. This information was later used for the segmentation of lungs. Korfiatis et al. [35] used a 3D multi-scale enhancement filter [36] to enhance vessels and vessel bifurcation points in CT images. This multi-scale filter can identify tubular structures, such as pulmonary vessels, by employing the eigenvalues of Hessian matrix at multiple scales. An expectation maximization algorithm is used for vessel segmentation by thresholding the response of voxels to the vessel enhancement filter at each scale. Wang et al. [37],[38] first used a threshold of -900 HU to segment air-component pixels and to identify the upper part of the trachea. They then used the identified trachea as a seed and traced the entire airways from an upper slice to a lower slice through the CT scan.

Segmentation of Lungs

Lung segmentation is the most important image segmentation technique used in CAD schemes for ILD. Most of the CAD schemes reviewed in this article have incorporated an automatic lung segmentation method. In early CAD schemes, edge gradient was frequently used to identify the boundaries of ribcages [25],[27],[28],[39]-[42]. The outer boundaries of lungs were first determined by identifying the minimum values of gray-level second derivative of the horizontal profiles, as described in [43] and [44]. And then the lung apex edges and diaphragm edges were detected by identifying an edge point at the location yielding the maximum edge gradient value in each horizontal line in an empirically determined range near the top and bottom of the chest radiography. The outer boundaries of lungs, apex edges, and diaphragm edges were finally combined to obtain the ribcage [44],[45].

Grey-level thresholding is another common technique used in the lung segmentation. However, this method becomes unreliable when the images contain high-attenuation disease patterns such as GGO, thus additional processing is needed to overcome the limitations of thresholding methods. Rudrapatna et al. [29] and Zrimec et al. [46] employed mathematic morphological operators to include high-attenuation abnormal lung parenchyma that were missed by thresholding-based segmentation method in CT images. Wang et al. [37],[38] first extracted initial lungs that contained the normal and mildly abnormal lung parenchyma in MDCT images by use of the thresholding technique. They then employed a texture analysis method to identify the abnormal lung parenchyma with severe disease patterns. Finally, they combined the initial lungs with the severely abnormal lungs to generate the final lung segmentation. Region growing-based lung segmentation methods were also widely used in CAD schemes for ILD [33],[34],[47],[48]. However, it might miss lungs with severely affected lung parenchyma. As an improvement, Arzhaeva et al. [49] used a multi-atlas-based method to improve the accuracy of lung segmentation obtained from region growing algorithm. In this method, pre-determined atlases of lungs were first registered to the segmented lung obtained with region growing algorithm, and then the atlases were utilized to guide the segmentation of lung with severe ILD in CT scans. Approaches based on active shape models algorithm [50],[51] and pattern classification algorithm [35] were also used for the segmentation of lungs in CAD schemes for ILD. Table 1 provides an overview of the automatic lung segmentation approaches reviewed in this article.

3.2. Determination of Region of Interest

A region of interest is a selected set of pixels (in 2D images) or voxels (in 3D images) within a medical image. The shape of ROI is typically a square in 2D images and a cube in 3D images.

Table 1. Lung segmentation methods used by CAD schemes for interstitial lung disease

Researchers	Method and Description	Dimensions	Imaging Modality
Hall et al.[27]	Analysis of intensity profile along x and y directions	2D	CRX
Powell et al.[44] Katsuragawa et al.[25],[41],[42] Morishita et al.[39] Monnier-Cholley et al.[40] Ishida et al.[28]	Analysis of profiles of grayscale second derivatives along x and y direction	2D	CRX
van Ginneken et al.[50]	Active shape models	2D	CRX
Shyu et al.[87] Uppaluri et al.[73]-[75]	Thresholding followed by a connected component labeling	3D	CT
Heitmann et al.[81],[90] Uchiyama et al.[52] Zavaletta et al.[31]	Thresholding	-	CT
Wang et al.[37],[38]	Thresholding followed by texture analysis-based refinement	3D	CT
Zrimec et al.[46]	Adaptive thresholding refined by use of mathematic morphology and active contour models	-	CT
Rudrapatna et al.[29]	Thresholding and mathematic morphology	-	CT
Marten et al.[33],[34]	Region growing	3D	CT
Arzhaeva et al.[68]	Region growing followed by an atlas-based approach for segmenting severely affected lung parenchyma	3D	CT
Park et al.[48],[77]	Adaptive thresholding and region growing	3D	CT
Korfiatis et al.[35]	Texture analysis-based lung parenchyma classification	3D	CT

-: Unknown.
CR: Chest Radiography.

Table 2. Summary of features, classification methods, and validation methods employed in CAD schemes for interstitial lung disease

Researchers	Imaging Modality	Disease Patterns	Features	Classifier	Validation Method	Performance Level
Katsuragawa et al.[25],[64],[65]	CRX	Nodular, reticular, and honeycombing	Fourier transform-based power spectrum	Rule-based	-	Az > 0.900
Katsuragawa et al.[41]	CRX	Nodular and reticular	Area of area components and length of linear components	Rule-based	-	Visual assessment
Katsuragawa et al.[21],[42]	CRX	Nodular and reticular	Fourier transform-based power spectrum, area of area components and length of linear components	Rule-based and ANN	-	Sen = 93.0%, Spe = 90.0%
Asada et al.[57]	CRX	Nine diseases: Sarcoidosis, silicosis, etc.	Clinical information and radiographic findings	ANN*	Jackknife	Az = 0.967
Chen et al.[43]	CRX	-	Fourier transform-based power spectrum	Rule-based	-	Az = 0.960
Kido et al.[55],[63]	CRX	Nodular and reticular	Laplacian-Gaussian filter features, length of linear elements, fractal features	Rule-based	-	The difference between feature values of normal and abnormal ROIs are statistically significant (t-test)
Kido et al.[66],[67]	CRX	-	Laplacian-Gaussian filter features, length of linear elements, fractal features	Rule-based		Az = 0.94 for mild ILD, Az = 0.98 for severe ILD
Monnier-Cholley et al.[40]	CRX	Eleven diseases: Sarcoidosis, silicosis, etc.	Fourier transform-based power spectrum	Rule-based	-	Sen = 92.0%, Spe = 90.0%
Morishita et al.[39]	CRX	-	Fourier transform-based power spectrum	Rule-based	-	Sen** = 90.0%, Spe*** = 89.0%
Floyd Jr. et al.[89]	CRX	Nodular	Fourier transform-based power spectrum	By observer	-	Az = 0.900
Ishida et al.[28]	CRX	-	Area of area components and length of linear components	Rule-based + ANN	-	Sen = 95.0%, Spe = 96.0%
Ashizawa et al.[58],[59]	CRX	Eleven diseases: Sarcoidosis, silicosis, etc.	Clinical information and radiographic findings	ANN	-	Az = 0.947, Sen = 0.890, Spe = 0.890
van Ginneken et al.[50]	CRX	-	Gaussian derivative filter features	KNN****	-	Az = 0.986
Abe et al.[56],[60]	CRX	Eleven diseases: Sarcoidosis, silicosis, etc.	Clinical information and radiographic findings	ANN	-	Az = 0.850
Arzhaeva et al.[49]	CRX	Focal pulmonary opacity, reticular, nodular, changes with increased density, and disease with cystic changes	Histogram features, x, y coordinate to the center of mass	LDA†, SVM††	4-fold cross-validation	Az = 0.780

Table 2. (Continued)

Researchers	Imaging Modality	Disease Patterns	Features	Classifier	Validation Method	Performance Level
Grenier et al.[72]	CT	Eight diseases: Sarcoidosis, silicosis, etc.	Clinical information, blood analysis features, and biologic features	Bayesian	-	Sen = 80.0%
Delorme et al.[69]	CT	GGO, intralobular fibrosis, and emphysema	Histogram features, run-length matrix, and co-occurrence matrix	LDA	2-fold cross-validation	Sen = 68.7-80.7%
Heitmann et al.[81],[90]	CT	GGO	Original pixel value, histogram features, co-occcurrence matrix,	ANN	-	Accuracy = 75.8%
Uppaluri et al.[73]-[75]	CT	Honeycombing, nodular, and emphysema	Histogram features, run-length matrix, coocurrence matrix, and fractal features	Bayesian	2-fold cross-validation	Sen = 93.5%
Shyu et al.[87]	CT	Reticular, nodular, high opacities, and low opacities	Histogram features, Sobel filter features, co-occurrence matrix	Multi-dimensional hashing-based classifi-cation	-	Sen =72.0%
Chabat et al.[30]	CT	Panlobular emphysema, centrilobular emphysema, and constructive obliterative bronchilolitis	Histogram features, co-occurrence matrix, and acquisition-length features	Bayesian	-	With probability thresholding: Sen = 60.3%, Spe = 86.7% Without probability thresholding: Sen = 73.6%, Spe = 91.2%
Sluimer et al.[83]	CT	-	Histogram features	KNN	4-fold cross-validation	Az = 0.862
Uchiyama et al.[52]	CT	GGO, reticular, nodular, honeycombing, emphysema, and consolidation	Histogram features, air density components, and mathematic morphological top-hat features	ANN	-	Sen = 97.4%, Spe = 88.0%
Fukushima et al.[61]	CT	Eleven diseases: Sarcoidosis, silicosis, etc.	Clinical information and radiographic findings	ANN	Leave-one-out	Az = 95.6%
Rudrapatna et al.[29]	CT	Pleural plaque and diffuse pleural thickening	Knowledge-based anatomical and pathological features	Bayesian	-	Sen = 81.0%, Spe = 95.0%, Accuracy = 94.0%
Xu et al.[76]	CT	GGO, honeycombing, and emphysema	Histogram features, run-length matrix, co-occurrence matrix, and fractal features	Bayesian and SVM	10-fold cross-validation	Bayesian: Sen = 87.6%, Spe = 96.2% SVM: Sen = 84.8%, Spe = 96.0%

Researchers	Imaging Modality	Disease Patterns	Features	Classifier	Validation Method	Performance Level
Depeursinge et al.[82]	CT	GGO, reticular, nodular, and emphysema	Histogram features and wavelet transform-based features	KNN	Leave-one-out	Accuracy = 92.5%
Fetita et al.[47]	CT	GGO, reticular, honeycombing, and emphysema	3D mathematic morphological features	Rule-based	-	Sen = 90.7%, Spe = 88.5%
Zavaletta et al.[31]	CT	GGO, honeycombing, reticular, and emphysema	Histogram features	KNN	10-fold cross-validation	Sen = 93.0%, Spe = 93.8%, Az = 0.931
Zrimec et al.[46]	CT	Honeycombing	Histogram features, co-occurrence matrix	Bayesian	10-fold cross-validation	Sen = 97.4%, Spe = 85.5.0%
Boehm et al.[70]	CT	Reticular and emphysema	Histogram features and topological features	LDA	Leave-one-out	Fibrosis: Az =0.87 Emphysema: Az = 0.96
Park et al.[77]	CT	GGO, reticular, honeycombing, emphysema, and consolidation	Histogram features, grey-level gradient features, run-length matrix, co-occurrence matrix, mathematic morphological top-hat features, and cluster of low attenuation area	SVM	5-fold cross-validation	Accuracy = 89.0%
Wang et al.[37]	CT	GGO, reticular, and honeycombing	Histogram features, run-length matrix, and co-occurrence matrix	QDA†††	Leave-one-out	Sen = 86.0%, Spe = 90.0%
Korfiatis et al.[35]	CT	GGO and reticular	Co-ocurrence matrix	KNN	10-fold cross-validation	GGO: Sen = 63.8, Spe = 63.9% Reticular: Sen = 94.2%, Spe = 85.3%
Arzhaeva et al.[68]	CT	-	Grey-level first- and second- derivative features	LDA, KNNandLDA	-	LDA: Accuracy = 76.1% k-NN and LDA: Accuracy = 79.5%
Kim et al.[78]	CT	GGO, reticular, and honeycombing	Co-occurrence matrix	SVM	-	Sen = 94.4%, Spe = 94.7%
Park et al.[48]	CT	-	Histogram features, run-length matrix	ANN	-	Sen = 80.0%, Spe = 85.7%, Az = 0.884

-: Unknown. CRX: Chest Radiography. *ANN: Artificial neural network. **Sen: Sensitivity. ***Spe: Specificity. ****KNN: K nearest neighbor. †LDA: Linear discrimination classifier. ††SVM: Supporting vector machine. †††QDA: Quadratic discrimination classifier.

The determination of proper size and location of ROIs is very important for accurate and reliable diagnosis of ILD. Manual ROI selection was commonly used in early CAD schemes. However, with the quick development and widespread use of high-resolution CT in the studies of ILD, manual ROI selection is becoming more and more impractical.

In Wang et al.'s study [37], contiguous 3D ROIs were automatically generated over the segmented lung regions with a fixed size and fixed distance between adjacent ROIs. This strategy was commonly used in many studies of CAD schemes for ILD [25],[27],[28],[39]-[42],[46],[37],[48],[49],[52] to automatically generate ROIs.

Chen et al. [43] proposed an automated method for selecting ROIs in the peripheral regions of segmented lungs in chest radiography, because ILD tends to be more clearly visible in the peripheral lung regions. The upper boundaries of the peripheral regions of both left and right lungs were determined at a distance below the lung apex that was equal to 20% of the height of the lung. The lower boundaries of the peripheral regions were determined at the top of the diaphragm. The outer boundaries were determined at 96% of the distance from the midline of the radiography images to the ribcage edge. The inner boundaries of peripheral regions were determined adaptively to decrease the width of the peripheral regions in the lower part of lungs to exclude large pulmonary vessels. Square ROIs were then evenly created in the identified peripheral lung regions. This strategy has been extensively used in the chest radiography-based CAD schemes. van Ginneken et al. [50] proposed a unique method to divide the segmented lung regions in chest radiography into ROIs with an identical area. As a first step, each lung was divided into an upper, middle, and lower part. Each of the three parts was subdivided into four regions of equal area, by first dividing it vertically and then horizontally.

Therefore, each lung was divided into 12 ROIs. Rudrapatna et al. [29] employed anatomical knowledge regarding disease patterns to define ROIs using a region growing technique along the lung boundaries.

3.3. Feature Extraction and Analysis

Abnormal lung parenchyma affected by ILD is usually characterized in appearance by unique texture-like patterns. Thus, disease patterns were detected and classified by use of texture features such as raw pixel values, gray-level histogram statistics (mean, standard deviation, skewness, kurtosis of intensity within a ROI), Fourier transform-based power spectrum [25], and features from run-length matrix [53] and co-occurrence matrix [54]. Edge detection filters, such as Sobel filters and Laplacian-Gaussian filters, were often used to extract the linear structures.These filters can enhance elements of linear opacities, such as reticular and honeycombing. In addition, geometry features have been employed widely in the identification of ILD.

The commonly used geometry features included circularity, compactness, mathematic morphological top-hat features [52], and fractal features, which was obtained using a box-counting method [55]. In some CAD schemes, clinical information and radiographic findings, such as age, sex, temperature, pleural effusions, and heart size were included [14],[56]-[61]. Table II lists the features commonly used in CAD schemes for ILD.

3.4. Pattern Classification

The last step of a CAD scheme of ILD is the classification of ROIs. The classification approaches are composed of two modules, i.e., a classifier algorithm for classifying disease patterns and a validation method for assessing performance level of CAD schemes. [62].

Rule-based classifier [21],[25],[28],[39]-[43],[47],[55],[63]-[67] was used in early CAD schemes for ILD. Later, more powerful statistical classification techniques, including linear discrimination analysis (LDA) [49],[68]-[71], quadratic discrimination analysis (QDA) [37], Bayesian classifier [29], [46], [72]-[76], support vector machine (SVM)[68],[76]-[80], artificial neural network (ANN), [28],[42],[48],[52],[56],[57]-[61],[81] and k-nearest neighbor (k-NN) cluster [35],[49],[50],[68],[82],[83] were employed in CAD schemes for ILD. Table II summarizes the pattern classification schemes used in published CAD schemes for ILD. It is apparent that powerful statistical classifiers were widely used in most recent CAD studies for ILD.

3.5. Validation of CAD Schemes

Validation Methods

In early CAD schemes, a resubstitution method was commonly used for evaluating the performance level of CAD schemes [84],[85]. Because the performance evaluated by the resubstitution method is subjected to bias, more reliable methods such as k-fold cross-validation [30],[31],[35],[46],[49],[62],[69],[73]-[77],[83],[86] were employed. The major advantage of k-fold cross-validation is that the bias in the estimated performance level of CAD schemes can be effectively reduced. Leave-one-out validation [37],[61],[70],[82] can be regarded as a special case of k-fold cross-validation. In practice, the choice of the number of folds depends on the size of the dataset. For a large dataset, the cross-validation with a small number of folds will be quite reliable, whereas, for a small dataset, a leave-one-out scheme may be more advantageous. Table 2 summarizes the validation methods for CAD schemes of ILD. It is apparent that cross-validation method was employed in nearly all recent CAD studies for ILD. This tendency indicates that the researchers are paying more attention to reducing bias in the estimated performance of CAD schemes.

Performance Metrics

Performance metrics such as the pair of sensitivity and specificity [21],[28],[29],[31],[35],[37],[39],[40],[42],[46],[47],[48],[52],[58],[59],[69],[72]-[76],[78],[87] and accuracy [29],[68],[77],[81],[82] are generally used:

$$\text{Sensitivity} = TP / (TP + FN)$$

$$\text{Specificity} = TN / (TN + FP)$$

$$\text{Accuracy} = (TP + TN) / (TP + FP + TN + FN)$$

where TP = number of true positives, FP = number of false positives, TN = number of true negatives, FN = number of false negatives. The area under receiver operating characteristic (ROC) curve (AUC or Az value) is another commonly used performance metric [25],[31],[43],[49],[50],[56]-[61],[64]-[67],[70],[83]. A ROC curve is a graphical plot of the sensitivity vs. (1 – specificity) for a classifier obtained by changing the threshold level for the output values of the classifier [88]. Because Az value is calculated based on a systematic analysis of sensitive and specificity of a CAD scheme, it provides a more comprehensive evaluation of performance levels of CAD schemes than a single pair of sensitivity and specificity or accuracy.

3.6. Observer Studies

Observer performance study is an important method for assessing the potential effect of CAD schemes on radiologists' performance. Ashizawa et al. [59] conducted an observer performance study to evaluate the effect of an ANN-based CAD scheme [58] on radiologists' diagnostic accuracy. Thirty-three chest radiographs with 11 ILD disease patterns (three radiographs per disease pattern) were selected for the observer study. Eight radiologists (four attending physicians and four residents) were recruited to conduct the observer study. Each radiologist first made diagnosis without, and then with, CAD output, which indicated the likelihood of each of the 11 possible diagnoses in each case. By use of the CAD output, the radiologists significantly improved their diagnostic accuracy for ILD from an average Az value of 0.826 to 0.911 ($p<0.0001$). Abe et al. [56] conducted a similar observer performance study for evaluating the usefulness of a CAD [60] in assisting radiologists to diagnose ILD in chest radiography. Five radiologists participated in the observer study. Thirty difficult chest radiographs with 11 ILD disease patterns were included in this study. The Az values for the CAD alone, radiologists without CAD, and radiologists with CAD were 0.85, 0.81, and 0.87, respectively. By use of CAD, radiologists improved their performance compared with CAD alone and radiologists without CAD.

Fukushima et al. [61] extended Ashizawa's work [58] to HRCT and conducted an observer performance study using a dataset of 45 HRCT scans that were selected from 130 HRCT scans. Each HRCT image was viewed by eight radiologists (four chest radiologists and four general radiologists) first without and then with CAD. The diagnostic performance (Az value) of chest radiologists and general radiologists was increased by use of CAD from 0.986 to 0.992 ($p = 0.071$) and 0.958 to 0.971 ($p < 0.001$), respectively. These observer performance studies demonstrated that CAD schemes could significantly improve the radiologists' ability in accurately diagnosing ILD in chest radiography and CT.

Conclusion

In this article, we introduced the basic knowledge of ILD and the concept of CAD schemes for ILD and reviewed the publications for the CAD schemes in terms of image segmentation, ROI determination, feature extraction and analysis, pattern classification, and observer performance study. Observer performance studies indicate that CAD schemes have potential to assist radiologists in the detection and diagnosis of ILD in chest radiography and CT. Although current CAD schemes for ILD are still in a relative early stage, it is expected that the CAD schemes will play an increasingly important role in the detection and diagnosis of ILD in the future.

References

[1] Colby, T., Swensen, S. *J. Thorac. Imaging* 11, 1 (1996).
[2] Reynolds, H. *Chest* 113, 192 (1998).
[3] British Thoracic Society Standards of Care Committee, *Thorax* 54, S1 (1999).
[4] Ryu, J., Olson, E., Midthun, D., Swensen, S. *Mayo Clin. Proc.* 1221 (2002).
[5] Schraufnagel, D. Breathing in America: Diseases, progress, and hope. *American Thoracic Society*, (2010).
[6] Coultas, D., Zumwalt, R., Black, W., Sobonya, R. *Am. J. Respir. Crit. Care. Med.* 150, 967 (1994).
[7] Sluimer, A. Schilham, Prokop, M., van Ginneken, B. *IEEE Trans. Med. Imaging* 25, 385 (2006).
[8] Austin, J., Müller, N., Friedman, P., Hansell, D., Naidich, D., Remy-Jardin, M., Webb, W., Zerhouni, E. *Radiology* 200, 327 (1996).
[9] Zompatori, M., Sverzellati, N., Poletti, V., Bna, C., Ormitti, F.,Spaggiari, E., Maffei, E. *Semin. Ultrasound CT MR* 26, 332 (2005).
[10] Collins, J. *Radiol. Clin. North Am.* 39, 1115 (2001).
[11] Hansell, D., Armstrong, P., Lynch, D., McAdams, H. *Philadelphia, Elsevier Mosby*, 69 (2005).
[12] Kazerooni, E. *AJR Am. J. Roentgenol.* 177, 501 (2001).
[13] Mathieson, J., Mayo, J., Staples, C., Müller, N. *Radiology* 171, 111 (1989).
[14] Grenier, P., Valeyre, D., Cluzel, P., Brauner, M., Lenoir, S., Chastang, C. *Radiology* 79, 123 (1991).
[15] Remy-Jardin, M., Remy, J., Deffontaines, C., Duhamel, A. *Radiology* 181, 157 (1991).
[16] Müller, N., Miller, R. Part 1. *Am. Rev. Respir. Dis.* 142, 1206 (1990).
[17] Müller, N., Miller, R. Part 2. *Am. Rev. Respir. Dis.* 142, 1440 (1990).
[18] Hoffman, E., McLennan, G. *Acad. Radiol.* 4, 758 (1997).
[19] Johkoh, T., Muller, N., Nakamura, H. *J. Thorac. Imaging* 17, 291 (2002).
[20] Kohz, P., Stabler, A., Beinert, T., Behr, J., Egge, T., Heuck, A., Reiser, M. *Radiology* 197, 539 (1995).
[21] Katsuragawa, S., Doi, K. *Comput. Med. Imaging Graph.* 31, 212 (2007).
[22] Doi, K. *Comput. Med. Imaging Graph.* 31, 198 (2007).
[23] Li, Q., Li, F., Suzuki, K., Shiraishi, J., Abe, H., Engelmann, R., Nie, Y., MacMahon, H., Doi, K. *Semin. Ultrasound CT MR* 26, 357 (2005).
[24] Li, Q. *Comput. Med. Imaging Graph.* 31, 248 (2007).
[25] Katsuragawa, S., Doi, K., MacMahon, H. *Med. Phys.*15, 311 (1988).
[26] Nishikawa, R., Digital Mammography, 85, Springer (2010).
[27] Hall, E., Crawford Jr, W., Roberts, F. *IEEE Trans. Biomed. Eng.* 22, 518 (1975).

[28] Ishida, T., Katsuragawa, S., Kobayashi, T., MacMahon, H., Doi, K. *Med. Phys.* 24, 915 (1997).
[29] Rudrapatna, M., Mai, V., Sowmya, A., Wilson, P. *Inf. Process. Med. Imaging* 19, 270 (2005).
[30] Chabat, F., Yang, G., Hansell, D. *Radiology* 228, 871 (2003).
[31] Zavaletta, V., Bartholmai, B., Robb, R. *Acad. Radiol.*14, 772 (2007).
[32] Sato, Y., Nakajima, S., Shiraga, N., Atsumi, H., Yoshida, S., Koller, T., Gerig, G., Kikinis, R. *Med. Image Anal.* 2, 143 (1998).
[33] Marten, K., Dicken, V., Kneitz, C., Hoehmann, M., Kenn, W., Hahn, D., Engelke, C. *Eur. J. Radiol.* 72, 278 (2009).
[34] Marten, K., Dicken, V., Kneitz, C., Hohmann, M., Kenn, W., Hahn, D., Engelke, C. *Eur. Radiol.* 19, 324 (2009).
[35] Korfiatis, P., Karahaliou, A., Kazantzi, A., Kalogeropoulou, C., Costaridou, L. *IEEE Trans. Inf. Technol. Biomed.* 14, 675 (2010).
[36] Zhou, C., Chan, H., Sahiner, B., Hadjiiski, L., Chughtai, A., Patel, S., Wei, J., Ge, J., Cascade, P., Kazerooni, E. *Med. Phys.* 34, 4567 (2007).
[37] Wang, J., Li, F., Doi, K., Li, Q. *Phys. Med. and Biol.* 54, 6881 (2009).
[38] Wang, J., Li, F., Li, Q. *Med. Phys.* 36, 4592 (2009).
[39] Morishita, J., Doi, K., Katsuragawa, S., Monnier-Cholley, L., MacMahon, H. *Med. Phys.* 22, 1515 (1995).
[40] Monnier-Cholley, L., MacMahon, H., Katsuragawa, S., Morishita, J., Doi, K. *Acad. Radiol.* 2, 455 (1995).
[41] Katsuragawa, S., Doi, K., MacMahon, H., Monnier-Cholley, L., Morishita, J., Ishida, T. *J. Digit. Imaging* 9, 137 (1996).
[42] Katsuragawa, S., Doi, K., MacMahon, H., Monnier-Cholley, L., Ishida, T., Kobayashi, T. *J Digit. Imaging* 10, 108 (1997).
[43] Chen, X., Doi, K., Katsuragawa, S., MacMahon, H. *Med. Phys.* 20, 975 (1993).
[44] Powell, G., Doi, K., Katsuragawa, S. *Med. Phys.* 15, 581 (1988).
[45] Nakamori, N., Doi, K., Sabeti, V., MacMahon, H. *Med. Phys.* 17, 342 (1990).
[46] Zrimec, T., Wong, J. *Stud. Health Technol. Inform.* 1129, 1324 (2007).
[47] Fetita, C., Chien, K., Brillet, P., Preteux, F., Grenier, P. "Diffuse parenchymal lung diseases: 3D automated detection in MDCT", 10th International Conference on Medical Image Computing and Computer Assisted Intervention, 29, Oct. 2007 (Brisbane, Australia) Part I, pp825-833.
[48] Park, S., Tan, J., Wang, X., Lederman, D., Leader, J., Kim, S., Zheng, B. *Phys. Med. and Biol.* 56, 1139 (2011).
[49] Arzhaeva, Y., Prokop, M., Tax, D., De Jong, P., Schaefer-Prokop, C., van Ginneken, B. *Med. Phys.* 34, 4798 (2007).
[50] van Ginneken, B., Katsuragawa, S., Romeny, B., Doi, K., Viergever, M. *IEEE Trans. Med. Imaging*, 21, 139 (2002).
[51] Cootes, T., Taylor, C., Cooper, D., Graham, J. *Comput. Vis. Image Underst.* 61, 38 (1995).
[52] Uchiyama, Y., Katsuragawa, S., Abe, H., Shiraishi, J., Li, F., Li, Q., Zhang, C., Suzuki, K., Doi, K. *Med. Phys.* 30, 2440 (2003).
[53] Galloway, M. *Comput. Graph. Imaging Process.* 4, 172 (1975).
[54] Haralick, R. *Proc. IEEE* 67, 786 (1975).

[55] Kido, S., Ikezoe, J., Naito, H., Masuike, M., Tamura, S., Kozuka, T. *Invest. Radiol.* 29, 172 (1994).
[56] Abe, H., Ashizawa, K., Li, F., Matsuyama, N., Fukushima, A., Shiraishi, J., MacMahon, H., Doi, K. *Acad. Radiol.* 11, 29 (2004).
[57] Asada, N., Doi, K., MacMahon, H., Montner, S., Giger, M., Abe, C., Wu, Y. *Radiology* 177, 857 (1990).
[58] Ashizawa, K., Ishida, T., MacMahon, H., Vyborny, C., Katsuragawa, S., Doi, K. *Acad. Radiol.* 6, 2 (1999).
[59] Ashizawa, K., MacMahon, H., Ishida, T., Nakamura, K., Vyborny, C., Katsuragawa, S., Doi, K. *AJR Am. J. Roentgenol.*, 172, 1311 (1999).
[60] Abe, H., Ashizawa, K., Katsuragawa, S., MacMahon, H., Doi, K. *Acad. Radiol.* 9, 13 (2002).
[61] Fukushima, Ashizawa, K., Yamaguchi, T., Matsuyama, N., Hayashi, H., Kida, I., Imufuku, Y., Egawa, A., Kimura, S., Nagaoki, K., Honda, S., Katsuragawa, S., Doi, K., Hayashi, K. *AJR Am. J. Roentgenol.* 183, 297 (2004).
[62] Duda, R., Hart, P., Stork, D. *Pattern Classification, 2nd Edition*, Wiley, New York (2000).
[63] Kido, S., Ikezoe, J., Naito, H., Tamura, S., Machi, S. *Radiographics* 15, 1457 (1995).
[64] Katsuragawa, S., Doi, K., MacMahon, H. *Med. Phys.* 16, 38 (1989).
[65] Katsuragawa, S., Doi, K., Nakamori, N., MacMahon, H. *Med. Phys.* 17, 72 (1990).
[66] Kido, S., Ikezoe, J., Tamura, S., Nakamura, H., Kuroda, C. *J. Digital Imaging* 10, 57 (1997).
[67] Kido, S., Tamura, S., Nakamura, H., Kuroda, C. *Comput. Med. Imaging Graph.* 23, 103 (1999).
[68] Arzhaeva, Y., Prokop, M., Murphy, K., van Rikxoort, E., de Jong, P., Gietema, H., Viergever, M., van Ginneken, B. *Med. Phys.* 37, 63 (2010).
[69] Delorme, S., Keller-Reichenbecher, M., Zuna, I., Schlegel, W., van Kaick, G. *Invest. Radiol.* 32, 566 (1997).
[70] Boehm, H., Fink, C., Attenberger, U., Becker, C., Behr, J., Reiser, M. *Eur. Radiol.* 18 2745 (2008).
[71] Fisher, R. *Ann. Eugen.*, 7, 179 (1936).
[72] Grenier, P., Chevret, S., Beigelman, C., Brauner, M., Chastang, C., Valeyre, D. *Radiology* 191, 383 (1994).
[73] Uppaluri, R., Mitsa, T., Sonka, M., Hoffman, E., McLennan, G. *Am. J. Respir. Crit. Care. Med.*, 156, 248 (1997).
[74] Uppaluri, R., Hoffman, E., Sonka, M., Hunninghake, G., McLennan, G. *Am. J. Respir. Crit. Care. Med.* 159, 519 (1999).
[75] Uppaluri, R., Hoffman, E., Sonka, M., Hartley, P., Hunninghake, G., McLennan, G. *Am. J. Respir. Crit. Care. Med.* 159, 648 (1999).
[76] Xu, Y., van Beek, E., Hwanjo, Y., Guo, J., McLennan, G., Hoffman, E. *Acad. Radiol.* 13, 969 (2006).
[77] Park, S., Seo, J., Kim, N., Park, S., Lee, Y., Park, B., Sung, Y., Lee, Y., Lee, J., Kang, S. *Korean J. Radiol.* 10, 455 (2009).
[78] Kim, H., Tashkin, D., Clements, P., Li, G., Brown, M., Elashoff, R., Gjertson, D., Abtin, F., Lynch, D., Strollo, D., Goldin, J. *Clin. Exp. Rheumatol.* 28, S26 (2010).
[79] Vapnik, V. *The nature of statistical learning theory*, Springer, New York, (1995).

[80] Vapnik, V. *Statitstical learning theory*, Wiley-Interscience, New York, (1998).
[81] Heitmann, K., Kauczor, H., Mildenberger, P., Uthmann, T., Perl, J., Thelen, M. *Eur. Radiol.* 7, 1463 (1997).
[82] Depeursinge, Sage, D., Hidki, A., Platon, A., Poletti, P., Unser, M., Muller, H. "Lung tissue classification using wavelet frames", 29th Annual International Conference of the IEEE Engineering in Medicine and Biology Society, 23, Aug. 2007 (Lyon, France) pp6259-6262.
[83] Sluimer, van Waes, P., Viergever, M., van Ginneken, B. *Med. Phys.* 30, 3081 (2003)
[84] Chan, H., Doi, K., Vyborny, C., Schmidt, R., Metz, C., Lam, K., Ogura, T., Wu, Y., MacMahon, H. *Invest. Radiol.*, 25, 1102 (1990).
[85] Kobayashi, T., Xu, X., MacMahon, H., Metz, C., Doi, K. *Radiology* 199, 843 (1996).
[86] Li, Q. *Comput. Med. Imaging Graph.* 31, 338 (2007).
[87] Shyu, C., Brodley, C., Kak, A., Kosaka, A., Aisen, A., Broderick, L. *Comput. Vis. Image Underst.* 75, 111 (1999).
[88] Metz, C. *Invest. Radiol.* 21, 720 (1986).
[89] Floyd Jr., C., Patz Jr., E., Lo, J., Vittitoe, N., Stambaugh, L. *AJR Am. J. Roentgenol.* 167, 1185 (1996).
[90] Kauczor, H., Heitmann, K., Heussel, C., Marwede, D., Uthmann, T., Thelen, M. *AJR Am. J. Roentgenol.* 175, 1329 (2000).

In: Medical Physics
Editor: Marcin Balcerzyk

ISBN: 978-1-62257-590-9

Chapter IX

Positron Emission Tomography of Small Animals Inspected from Biologist, Radiochemist and Engineer Points of View: What Each of Them Does Not Understand in the Other's Domain?

Marcin Balcerzyk[*]
National Center for Accelerators, University of Seville, Spain

Abstract

We are approaching physical limits of spatial resolution in new scanners for Positron Emission Tomography (PET) determined by positron range in tissues. On the other hand, Computed Tomography and Magnetic Resonance Imaging with much higher spatial resolution but less functional information is now being united in one PET/CT or PET/MRI scanner machine. While physicists and engineers develop equipment with higher and higher spatial resolution, the end users of tomography equipment very often are unable to achieve their goals of molecular imaging, analyze the information contained in the images received in several modalities, optimize experimental parameters to receive good kinetic and pixel-wise modeling and obtain the images of desired quality. On the other hand, some essential requirements of end users – biologists, pharmacologists or, medical doctors as uniformity of calibration, partial volume correction, need for speed in the reconstruction, etc. are not understood or not addressed properly by the designers of the equipment. Radio-chemists are adding new compounds to a list of PET tracers, and moreoften, some new isotopes, with more difficult to understand properties, which are not only summarized in halflife and positron range. This article tries to bridge these three

[*] E-mail address: mbalcerzyk@us.es.

worlds; explaining essential issues for each world in a language comprehensible to the other with a large number of examples from preclinical imaging of small animals.

1. Introduction

We are approaching physical limits of spatial resolution in new scanners for Positron Emission Tomography (PET) determined by positron range in tissues. On the other hand, Computed Tomography and Magnetic Resonance Imaging with much higher spatial resolution but less functional information is now being united in one PET/CT or PET/MRI scanner machine. While physicists and engineers develop equipment with higher and higher spatial resolution, the end users of tomography equipment very often are unable to achieve their goals of molecular imaging, analyze the information contained in the images received in several modalities, optimize experimental parameters to receive good kinetic and pixel-wise modeling and obtain the images of desired quality. On the other hand, some essential requirements of end users – biologists, pharmacologists or medical doctors as uniformity of calibration, partial volume correction, need for speed in the reconstruction, etc. are not understood or not addressed properly by the designers of the equipment. Radio-chemists are adding new compounds to a list of PET tracers, and more often, some new isotopes, with more difficult to understand properties, which are not only summarized in half-life and positron range. This article tries to bridge these three worlds; explaining essential issues for each world in a language comprehensible to the other with a large number of examples from preclinical imaging of small animals.

2. Engineer's World

2.1. Spatial Resolution in the Remote Positions of the Field of View

Spatial resolution of a microPET system deteriorates outside from the center of the field of view. Resolution is usually measured with a small activity (tens of kBq ^{22}Na) point source (i.e. less than 1 mm size). In an image, it results in an ellipsoid, which dimensions may be approximated in three orthogonal directions: axial one (along the axis of the scanner), radial one (along the radial axis perpendicular to the main axis of the system), and tangential one (which is perpendicular to the other two). The data of the resolution for two microPET systems are published in [1, 2]. Most microPET systems have between 40 and 150 mm of axial field of view, with usual of 120 mm. Radial field of view is about 80 – 100 mm diameter. The dimensions of a typical mouse of 32 g weight are about 100 mm length (excluding the tail) and about 23 mm in largest diameter at the end of the ribs. A typical rat of about 250 g is 200 mm long (excluding the tail) and about 50 mm largest diameter at the end of the ribs. For imaging of the full body rats, a multi-bed scan is used with some overlap to join the images. Typical heated beds have ∅70 mm for a rat, and ∅35 mm for a mouse.

Therefore, for both rat and mouse scan, we would need full length of a bed and about 63% of radial field of view for a rat and 32% for a mouse (80 mm radial FOV system) if an animal is positioned at the axis of the system. In some cases, two or even three mice are imaged simultaneously side by side or in triangular arrangement in a multi-bed system (Minerve, France). In such a system, we would need radial space similar to a rat. Sometimes for brain studies, animals are positioned head to head facing each other and they are positioned on the axis, or somewhat off axis, to get almost full body images.

Usually the axial and tangential resolutions are almost constant within the field of view. The one which is changing a lot is radial resolution at radial direction. For a system described in [2] at the center of the field of view, the volumetric resolution is about 5.4 μl, at the ribs of a mouse is about 7.2 μl, at the ribs of a rat about 9.0 μl. For the system described in [1] at the center, the volumetric resolution is about 4.1 μl, at the ribs of a mouse is about 5.4 μl, at the ribs of a rat about 8.1 μl. These resolutions are telling what the minimum volumes which may be detected are in PET images that means that a source of smaller volume will have the activity image distributed within resolution volume. For the linear resolution, usually the smallest value of the three, corrected for the finite size of the source at the center of the field of view called intrinsic resolution – for a system described in [1], it is 1.2 mm, and for the system described in [2], it is 1.4 mm. The main cause of deterioration of the radial resolution outside of the axis of the system is parallax error. Shortly, one can describe it in the following example. Typical crystals of the detector are 1.5×1.5×10 mm dimensions, with the long axis parallel to the radial direction. The rat is placed in the axis of microPET system. If positron emitting isotope is injected and absorbed in one of the ribs, and two annihilation γ are detected in a vertical pair of detectors, there is an error in registering radial coordinates of this line of response due to inclined crystals, as the crystals are facing 1.5×10 mm surfaces. If two annihilation γ are detected in a horizontal pair of detectors, perpendicular to the axis of the system, there is practically no parallax error, as the crystals are facing 1.5×1.5 mm surfaces.

A system described in [2] has 3.2 mm radial resolution at a radial distance of 42 mm. If the resolution 1.4 mm at the center is an intrinsic one, additional component at this radial distance is 2.88 mm from error propagation in quadrature. This FWHM is equal to 1.22 mm standard deviation. $6\sigma$is equal to 7.34 mm, and this corresponds exactly to a 10 mm face of the crystal projected on the radial axis. This system does not have depth of interaction (DOI) detection of the γ in the crystal. A system described in [1] has a newer version of 148 mm axial field of view and DOI detection and correction. The crystals in that system are tapered from 40×40 mm to 50×50 mm and 10 mm thickness. The system has ∅80 mm radial FOV and ∅105 mm bore size. Intrinsic resolution at the center of the FOV is 1.2 mm. Of course, DOI detection has some error in such a large crystal. The average error (FWHM) is 3.9 mm with minimum values at the center at 1.6 mm to maximum at the corners of 7.3 mm [3]. Calculating as in the previous paragraph, but now 3.9 mm FWHM error, we receive the radial resolution of 2.28 mm at 37 mm radial direction. Extrapolating to a radial extent of the mouse of 12 mm, we receive at this position 1.31 mm resolution and for the radial extent of a rat of 25 mm, we receive 1.69 mm resolution. So, volumetric resolution are 3.24 μlat the center, 3.5 μl at the side of the mouse body (12 mm radial) and 4.6 μl at the side of the rat body (25 mm). A typical test *in vivo* for resolution is a microPET/CT image of a full body mouse or rat with [^{18}F]NaF. Fluorine ion within 20 minutes of intravenous injection is incorporated into bones of the animal. Bones are excellently imaged in CT at a resolution much higher than PET. When images are fused (i.e. co-registered, meaning the same position of each anatomical element when both images are overlaid), one can see the fine details of skull, spine and ribs and estimate in situ resolution of the microPET system.

2.2. Partial Volume Effect

Finite resolution of the microPET system is responsible for blurring of the imaged activity. The effect is especially pronounced in small details of the image, for example, small nuclei in the brain, veins, and ventricles of the heart. Ventricles are additionally blurred by the myocardium movement. A simple rule of thumb says that for objects of sizes less than two times of system resolution, partial volume effect occurs [4]. The effect is especially important if quantification of the image is used, i.e. object size measurements or activity concentration measurements.

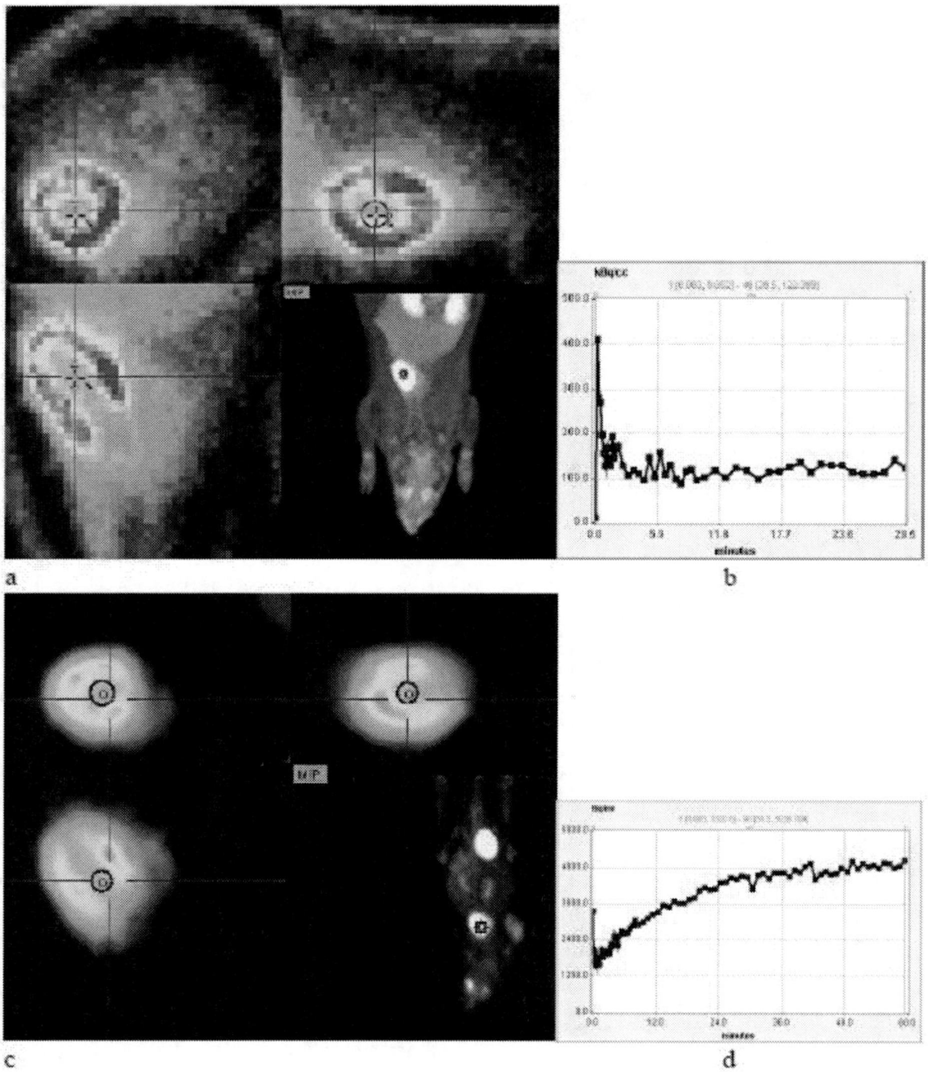

Figure 1.Partial volume effect in rat and mouse hearts[18F]FDG-PET image. (a). Image of rat heart. Three orthogonal projections and maximum intensity projection in lower right pane. (b) Image derived input function from left ventricle of the heart of acceptable quality - not corrected for activity of the sample of the blood taken at the end of the experiment.(c) Image of mouse heart. Three orthogonal projections and maximum intensity projection in lower right pane.(d) (b) Image derived input function from left ventricle of the heart of unacceptable quality - activity from myocardium is dominating.

The measured object appears larger in size than in reality especially for the object in the range of one resolution. More critical is the issue for peak activity measured and activity concentration. For example, for an uniformly loaded sphere of activity of a diameter of one resolution, the peak activity in the center is only 30% of the real one (so-called recovery coefficient). If the sphere has twice the resolution diameter, the recovered peak activity is about 80%. At the diameter of 3 resolutions, the recovered peak activity is close to 100%. The measurement of recovery coefficient is shown in [2] for a line source of varying diameter.

For the measurement of activity concentration in the image, one has to use a shape constructed from the pixels of the image. For the images where values in the pixels are showing concentrations of activity, the minimum volume is one pixel of the image. For the system described in [1], the image pixel size is 0.4 × 0.4 × 0.8 mm, which is 0.128 μl. The resolution in the center of the FOV is 1.5 × 2 ×1.8 mm (radial, axial, tangential FWHM), which gives 5.4 μl volumetric resolution. Therefore, all objects of linear dimension less than about 2 mm (ca. 5 μl, or 40 voxels) will suffer extended dimensions and blurring and the objects of linear dimensions less than about 5 mm (ca. 65 μl or 500 voxels) will have distorted activity concentration. The latter limitation is quite important for small tumor studies or studies of small nuclei in the brain which rely on quantification. One way to overcome the problem is to use small sampling volumes of several pixels only. As shown in [4], if we sample the sphere of diameter of two resolutions with a volume of interest of one resolution size, the recovered concentration of activity is only 55% of the real one. While the sampling is done with the volume of interest of only 0.4 of the resolution, the recovered concentration is 80% of the real one.

The difference becomes clear when we try to obtain image-derived input function from the left ventricle of the small animal. In cases of high resolution systems[2], one can obtain it for [^{18}F]FDG scan for a rat but not for a mouse, because the movement of the myocardium and its small thickness make the imaged activity "flow" to the neighboring area of the ventricle. The effect is illustrated in Figure 1.

2.3. PET/CT and PET/MRI Registration and Multimodal Imaging

Not often the laboratories have the luxury of having PET, CT and MRI small animal scanners with the transferable animal bed which allows taking of the images of the animal without movement of the animal with respect to the bed. PET/CT systems are increasingly popular, and some companies are developing PET/MRI systems already developed in the labs for simultaneous measurements [5].

However, in many cases, none of the PET, CT or MRI modalities give the full information about the studied case. We used a mouse model of thyroid cancer induced by prolonged drinking of water with perchlorate, which blocks iodine incorporation in thyroids and atfirst causes hyperplasia and then thyroid cancer. Iodine is transported actively by sodium-iodine-symporter (NIS), which, in rats, is expressed chiefly in the thyroids and stomach, but in mice also in salivary glands[6, 7]. For NIS imaging, we used [^{18}F]$NaBF_4$ new tracer [8]. In rats, as expected, it imaged thyroids, but in mice the imaged was dominated by salivary glands. Thyroids in mice and rats consist of two lobes (right and left) located on the sides of the larynx, posterior to hyoid bone. The estimated thyroids volume in healthy mouse is about 3.6 μl [8] which is at the limit of volumetric resolution of best microPET scanners.This organ is highly vascularized, but the vascularization is not uniform enough to delineate the full extent of the thyroids, to measure its volume and hence level of hyperplasia. One can easily approximate localization of thyroids in CT image – the larynx shows as air with HU around -1000, and hyoid bone a few mm anterior of the region. PET and CT images were co-registered using PMOD software [9]. Not having an MRI scanner, we used a provided T2 weighted image of the neck and brain region of another mouse done in another imaging canter, which was co-registered to CT image at the area of larynx and hyoid bone. In MRI images, one could locate the vascularized region close to the larynx, which was also visible. That allowed us to delineate part of the thyroids, which have high vasculature. The region was selected for NIS analysis via [^{18}F]$NaBF_4$ incorporation. With our nanoCT scanner from Bioscan we were not able to see vasculature with FenestraVC®contrast (ART Advanced Research Technologies Inc., Montréal, Canada) [10].

For a more precise measure of hyperplasia of the thyroids, we used parametric mapping with PMOD software, using image-derived input function from the left ventricle region of the heart of the mouse. [^{18}F]$NaBF_4$ does not incorporate in the myocardium, so the heart is the proper region for taking image-derived input function, as the images were taken in dynamic mode. The starting region of interest was the already delineated in MRI image. Two tissue compartment models were used to model the kinetic of [^{18}F]$NaBF_4$ incorporation. This procedure allowed us to prepare a parametric map of K_1 parameter and finally estimate volume of thyroids.

3. Radiochemist World

3.1. Radioisotopes Other Than Fluor-18 and Carbon-11

Most of the PET radiotracers use ^{18}F (110 min half-life) or ^{11}C (20 min half-life) as source of positrons which are then detected in PET camera detectors. Less used are ^{13}N (10 min half-life) and ^{15}O (2 min half-life). The principal reason for the use of carbon isotope ^{11}C is that most molecules, which are thought to be used in PET, contain at least one carbon atom. More specifically, the best carbon group to substitute with ^{11}C is some methyl (CH_3-) group in the compound, as the radioactive reaction precursor is [^{11}C] methyl iodide - $^{11}CH_3I$.

On the other hand,^{18}F when placed in radioactive tracer usually substitutes –OH group of the original compound, and changes the chemical properties substantially. ^{18}F having almost six times longer half-life than ^{11}C and lower positron maximum energy is a preferred radioisotope as it allows long range transport of synthetized radiotracers.

One of the less frequently used isotopes which is used in PET radiotracers is ^{64}Cu. Let us take it as an example of analysis of properties relevant for PET imaging. There are many tables of isotopes available, but only some contain detailed data on beta decay that permit analysis of isotope properties[1]. ^{64}Cu of 12.7 h half-life has two modes of decay – β^- and β^+. – β^- has 39% branching ratio (i.e. fraction of this mode of decay) and β^+61% branching. In β^-

[1]For subsequent analysis.we use the data from the Lund table of isotopes, http://nucleardata.nuclear.lu.se/nucleardata/toi/ and Isotope Explorer v 2.23, program (http://ie.lbl.gov/isoexpl/isoexpl.htm), based on the same data.

decay, the reaction that takes place in the nucleus is ${}^{A}_{Z}X \rightarrow {}^{A}_{Z+1}X + e^{-} + \bar{\nu}_{e}$ (i.e. neutron in the nucleus decays to proton, electron and anti-neutrino). Proton is captured in the nucleus and therefore, ${}^{64}Cu$ transmutes to ${}^{64}Zn$.In β^{+} decay, if the energy of unstable nucleus is higher by a minimum of 1022 keV from a daughter nucleus, the following reaction takes place: ${}^{A}_{Z}X \rightarrow {}^{A}_{Z-1}X + e^{+} + \nu_{e}$. ${}^{64}Cu$ transmutes to ${}^{64}Ni$ – branching ratio for this reaction is 43.1% electron capture without positron emission[2], 0.47% fraction of positron emission accompanied by γ of 1346 keV,17.4% positron emission without accompanying γ(see Figure 2), and 39% of electron capture without positron emission where ${}^{64}Cu$ transmutes to ${}^{64}Zn$(see Figure 3).

Therefore, if the radioisotope is marked with ${}^{64}Cu$, in the PET system per each 100 decays of ${}^{64}Cu$ will appear 0.47 γ of 1347 keV, and 2×17.4 γ of 511 keV resulting from anihilation of the positron.The rest of the decays will result in X-rays characteristics for K_{α}, K_{β}, etc from Ni and Zn, respectively.

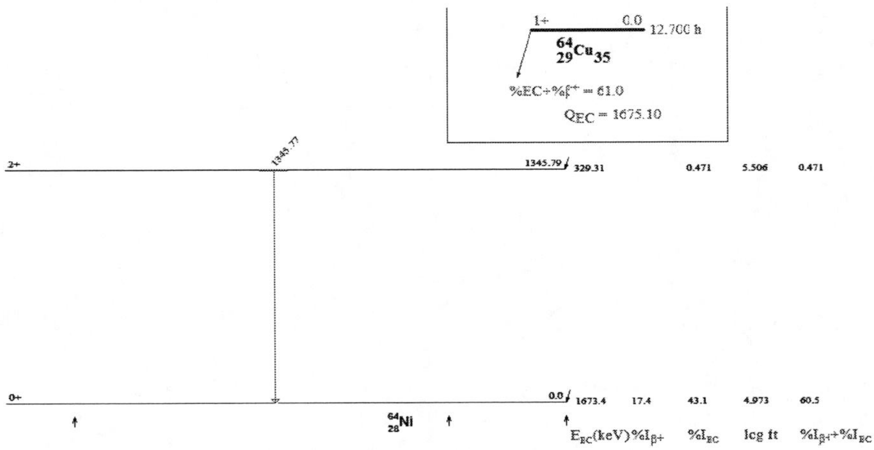

Figure 2. Scheme of β^{+} decay of ${}^{64}Cu$.

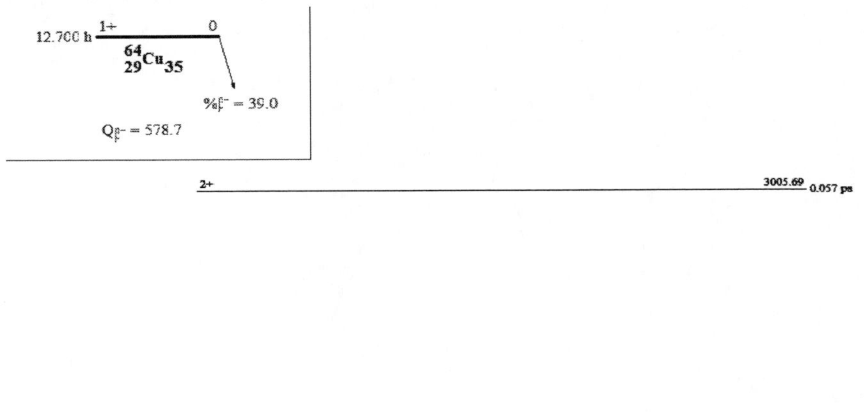

Figure 3. Scheme of β^{-} decay of ${}^{64}Cu$.

[2] $1s^2$elecrons have the highest probability being present at the nucleus, as the wave function of these electron is non-zero at the center of atom.

Neutrinos will not be detected. The PET system will detect γ of 511 keV, and for systems without upper limit of detection window, also γ of 1347 keV. Even if the system has limitations at a higher energy of the window active, for example, of 650 keV, the system will detect some Compton scattered events coming from absorption of γ of 1347 keV in detector crystals. Such scattered Compton events cannot be distinguished from photoelectrically absorbed γ of 511 keV, as the resolution of PET detectors is not sufficient for that. These would be random coincidences.

Let us have a look at other high weight isotopes of ^{124}I, ^{44}Sc and ^{68}Ga which are sometimes used in PET. The properties are summarized in Figure 4, Figure 5, Figure 6 and Table 1.

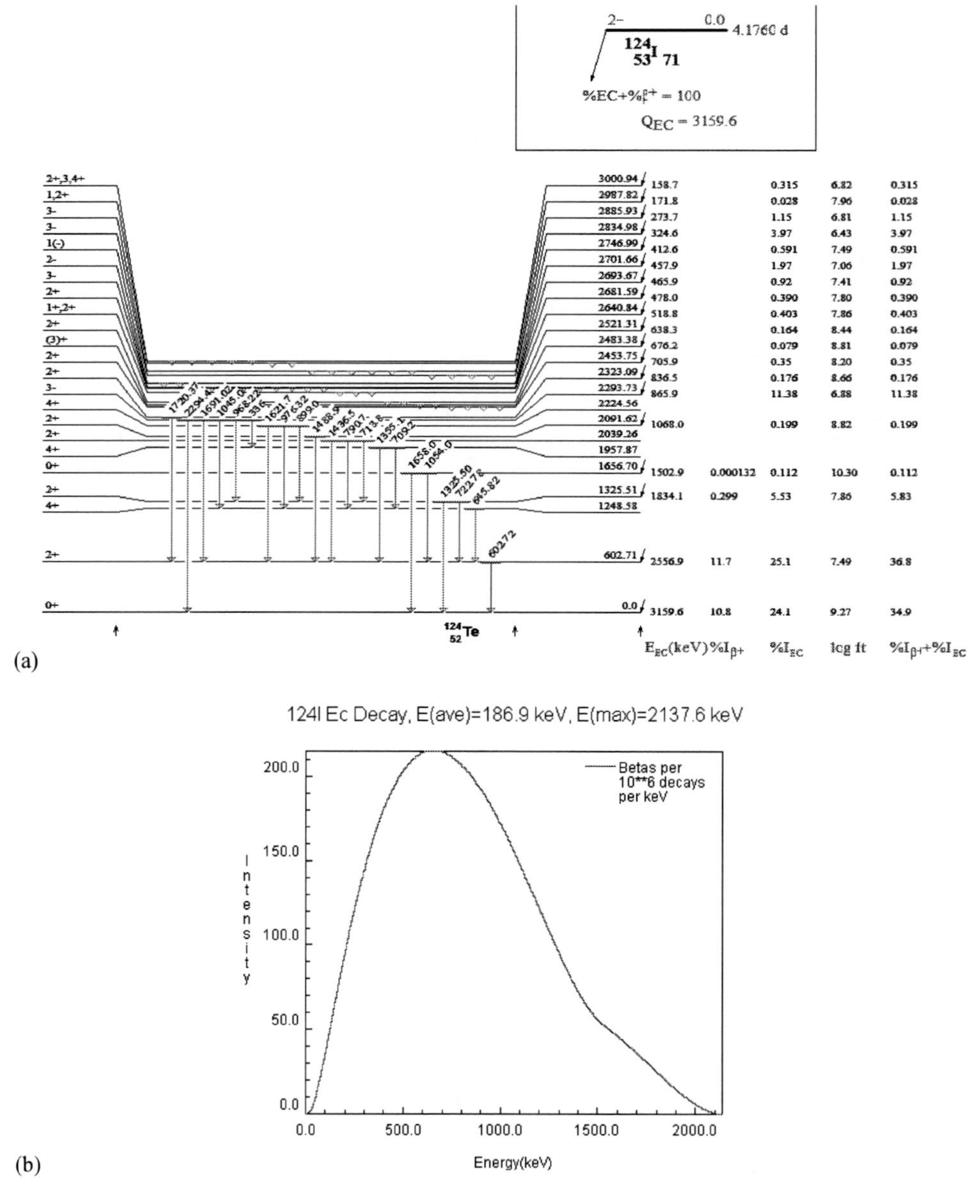

Figure 4. Level scheme (a) and beta spectrum (b) of β^+ decay of ^{124}I. Correct average positron energy is 822 keV.

The additional γ lines influence the detector performance. Some systems do not actually have high energy limits in energy windows and register the additional γ. In these, where the upper limit of the energy window is imposed, we would observe the photoelectric peak, if γ line lies within energy window, or a part of it which is blurred by the resolution of the system. If the peak lies sufficiently higher above the upper limit of energy window, we will see only Compton scattered events. These γ lines origin from the states of very short half-life usually (ps range) and therefore they will be detected in coincidence with 511 keV from positron annihilation. One immediate application can be considered – the use of triple coincidence to reduce random events rate. As the third particle is involved, the detection efficiency will be lower, but on the other hand, for the human PET systems with high random rate, reaching 50%, it may be beneficial. Such modification requires preparation of different electronics.

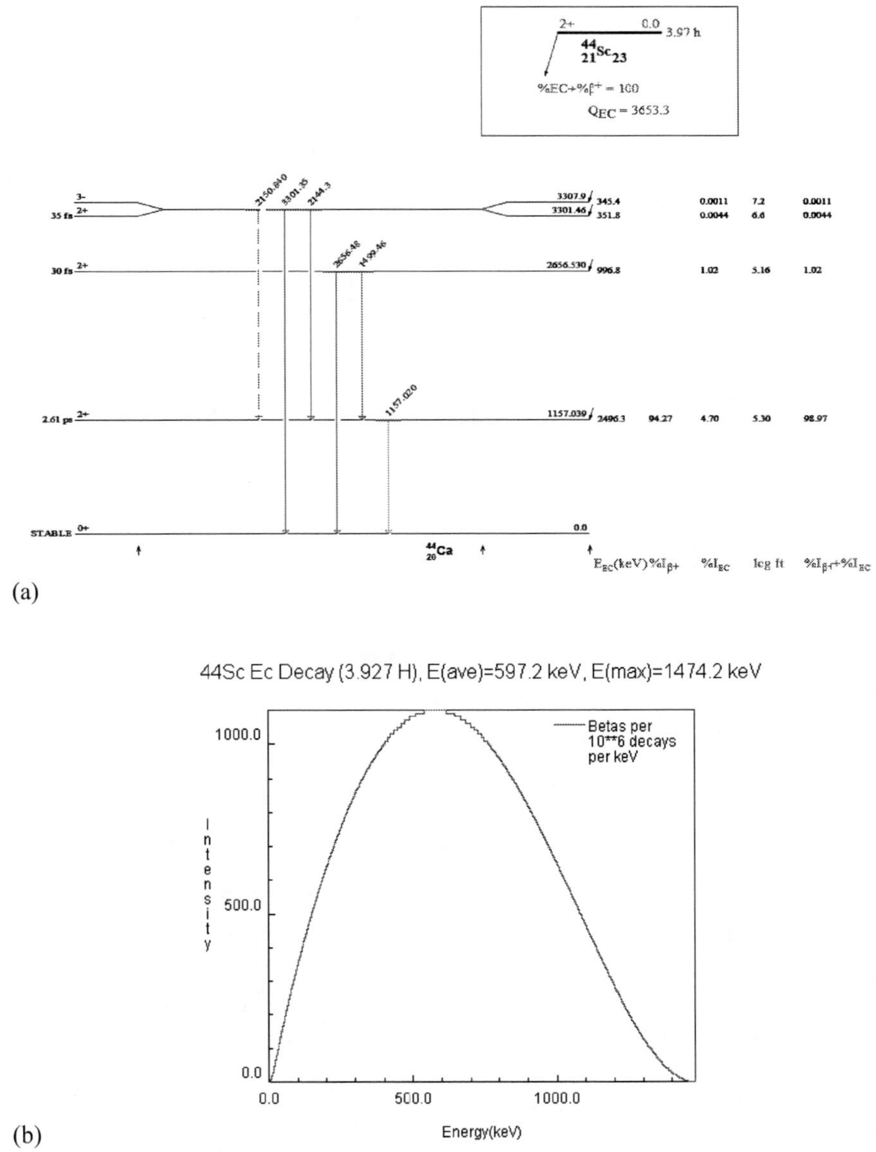

Figure 5. Level scheme (a) and beta spectrum (b) of β^+ decay of ^{44}Sc.

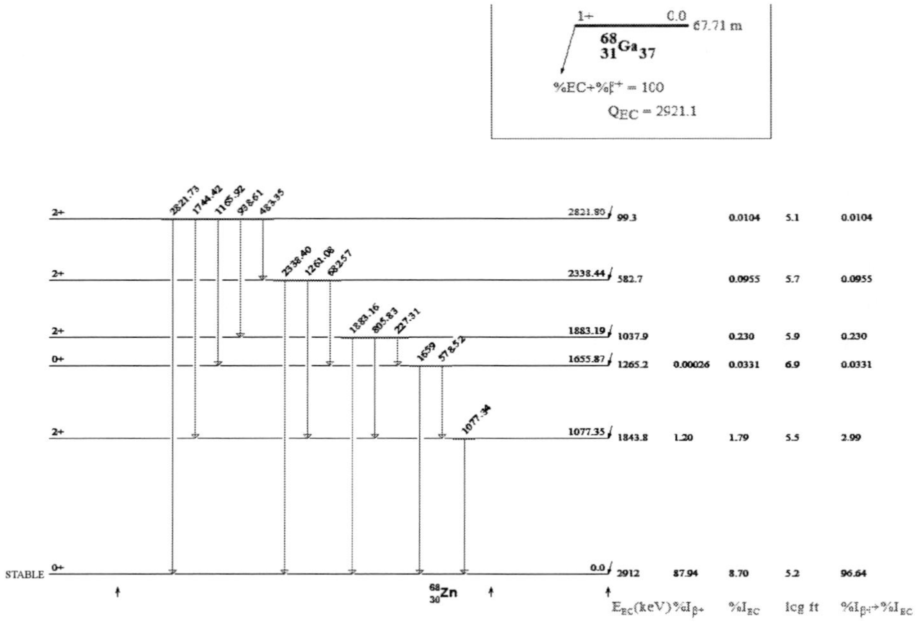

(a)

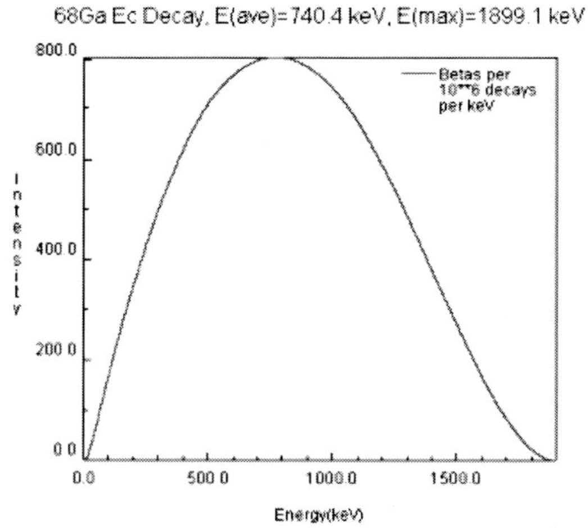

(b)

Figure 6. Level scheme (a) and beta spectrum (b) of β^{+68}Ga decay.

For ^{44}Sc, an interesting application was proposed to use Compton telescope from additional γ which would be pointing to the decay point of the PET isotope atom [11]. Shortly, additional γ is assumed to leave the patient´s body without interaction and being Compton scattered in a tank of liquid argon. Liquid argon is known to be a good scintillator, and such a detector can register energy and direction of the incoming γ, as the first interaction produces ionization shower resembling a straight thunderbolt. If a tank of liquid argon is mounted at the side of PET detector ring, almost half of such additional γ may be detected. From detected parameters of Compton interaction, one can find a line forming a Compton cone slant height line pinned to the interaction point in the argon detector. Such a cone is defined by its tip, direction and angle.

Table 1. Properties of decays of isotopesless frequently used inPET

	Property	^{124}I	^{44}Sc	^{68}Ga
1.	Half-life	4.176 d	3.97 h	67.71 min
2.	% decay β^+ to ground state (max positron energy)	10.8% (2138 keV)	-	88% (1899 keV)
3.	% decay β^+ to all excited states(max positron energy)	12% (1535keV)	94.27% (1472 keV)	1.2% (822 keV)
4.	γ accompanying positron, keV	602.7 (to ground state of daughter stable isotope ^{124}Te); 722.8 ; 1325; 602.7	1157	1077 (to ground state of daughter stable ^{68}Zn);

Energy and direction of detected $\gamma$are measured in a liquid argon detector. From known energy of a γ emitted from PET isotope before Compton scattering and measured energy, one can find the angle of the Compton cone. A cross-section of this cone surface and conventional line of response detected by a pair of PET detectors indicates a point where ^{44}Sc isotope atom was present. More precisely, a Compton cone is defined from the emission of additional γ from isotope, but line of response is defined by the annihilation point of the positron which after emission from the ^{44}Sc is scattered in the patient before annihilation. The difference is due to positron diffusion.

3.2. Positron Acolinearity, Mean and Maximum Range

Except for the resolution limitations of the PET scanner resulting from design, there are two physical limitations of the resolution.

1. First is due to non-perfect co-linearity of the two γ resulting from positronannihilation, due to some momentum being carried by neutrino, which is more important for large diameter systems.
2. Second is due to a finite range of the positron in the patient, and that only after this scattering, the annihilation happens when the positron loses all of its kinetic energy. The simulation of this effect for several PET isotopes was done in [12],[13, 14] and measured in [15]

The effect of acolinearity is shown for cylindrical PET system of $2r = 80$ cm diameter in Figure 7 for ^{18}F. For other isotopes, the acolinearity is larger, as the maximum positron energy and hence the positron momentum would be larger. As the acolinearityvalues $\Delta\varphi$ are shown in degrees, the value is practically independent of the system type, i.e. bore diameter, which means that the positioning error at the detectors scales as $\frac{r}{2}\Delta\varphi$[16]. Full width at half maximum (FWHM) of the curve is about 0.58deg, so the error for the 80 cm PET system is about 2.04 mm. More exact measurements were done by Shibuya [16] and led to an image blurring estimation of $(0.00243 \pm 0.00014) \times 2r$. This estimation for the mentioned scanner leads to a blurring estimate of 1.94 mm, which is close to current resolution of the human scanners. Using the blurring value for human measurement to a small animal scanner of $2r$ = 105 mm and resolution of 1.2 mm, we receive 0.26 mm blurring, which is much less than this resolution, but already contributing to it. Positron acolinearity for several isotopes is summarized in Table 2 in units of root mean square (RMS) angle.

Positron range contribution to blurring is more difficult to estimate. Its empirical distribution function $q(r)$at distance $r$in water from positron emission point is $q(r) = Ae^{(-r/B)} + (1 - A)e^{(-r/C)}$, where A,B,C are constants [15]. Even though the positron has finite energy and mass, in such distribution, there is no clear maximum distance defined, as for protons or heavy ions (Bragg peak). The blurring caused by positron diffusion may be characterized by FWHM of this distribution, full width at tenth of the maximum (FWTM) and root mean square distribution width. In Figure 8, we show calculated positron distance for ^{11}C, ^{18}F and ^{68}Ga according to [15].

Positron distance for various geometries of the source and for various materials may be simulated for PET systems using Gate program [17]. For the ^{124}I ∅1 mm line source in a cylindrical water tank, the positron distance is shown in Figure 9. For this source, the RMS is 1.303 mm.

For positron range blurring we would therefore suggest to use RMS (or variance, σ) of the distribution, and for maximum range, we would suggest 3σ as FWHM and FWTM values heavily rely on the precision of the value at zero distance. A set of values compiled from [12, 15] and calculated with Gate [17] is listed in Table 3.

Table 2.Positron acolinearity for several positron sources

Source	^{18}F	^{11}C	^{15}O	^{124}I	^{44}Sc
Acolinearity, deg	0.2484	0.2497	0.2505	0.2439	0.2467

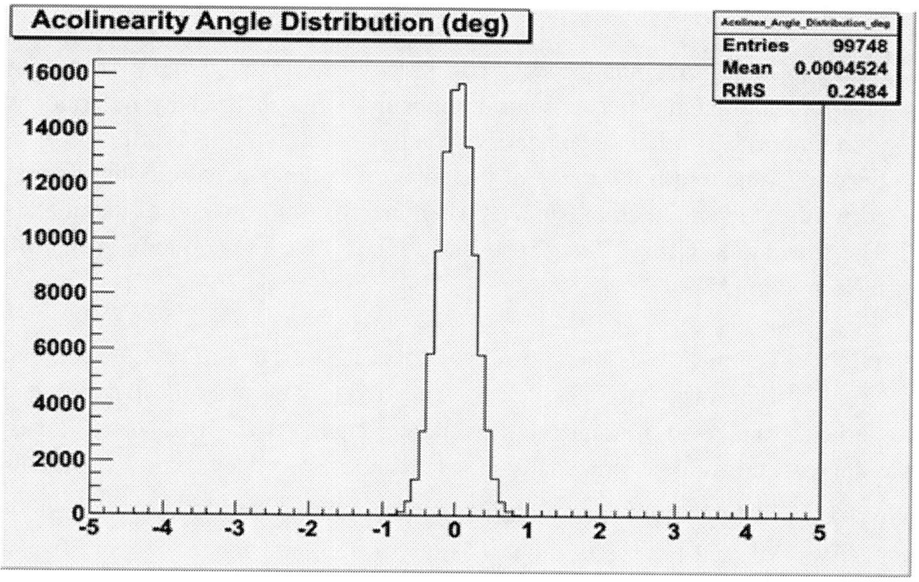

Figure 7. Acolinearity angle distribution(in degrees) for cylindrical PET system. Result from Gate simulation program version 5[17] and benchmarkPET systemfor ^{18}F source.

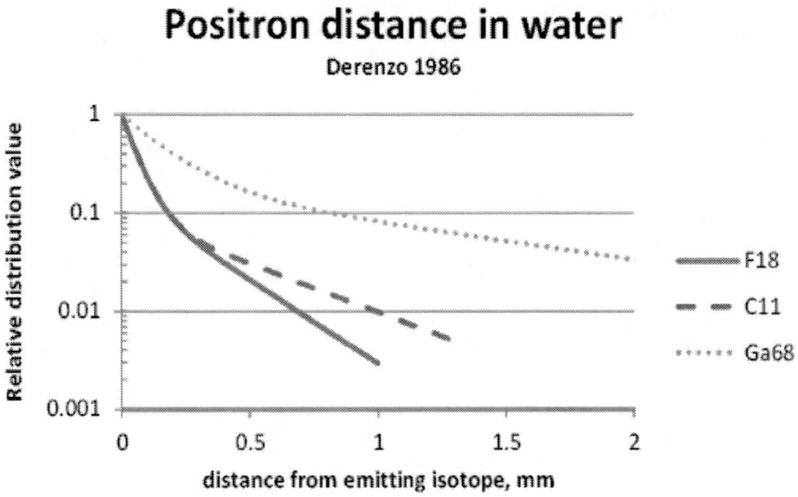

(a)

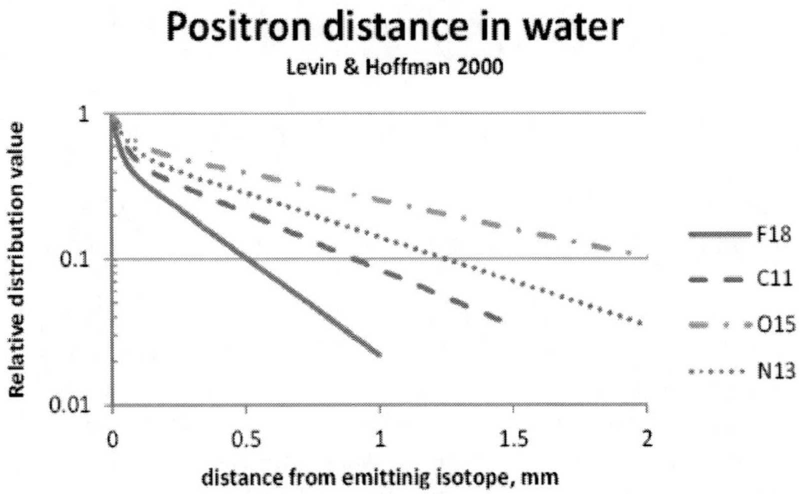

(b)

Figure 8.Positron distance in water for (a): ^{18}F, ^{11}C and ^{68}Gapoint source isotopes calculated according to empirical expression from [15], (b) ^{18}F, ^{11}C, ^{15}O and ^{13}N calculated according to empirical expression from [14].

Table 3. Image blurring component from point source and positron range in water

Isotope	Blurring due to e^+ range, RMS, mm	e^+ range, 3×RMS, mm
^{18}F	0.23	0.69
^{11}C	0.39	1.17
^{15}O	1.453	4.36
^{68}Ga	1.2	3.6
^{124}I*	1.303	3.91

* for∅1 mm line source.

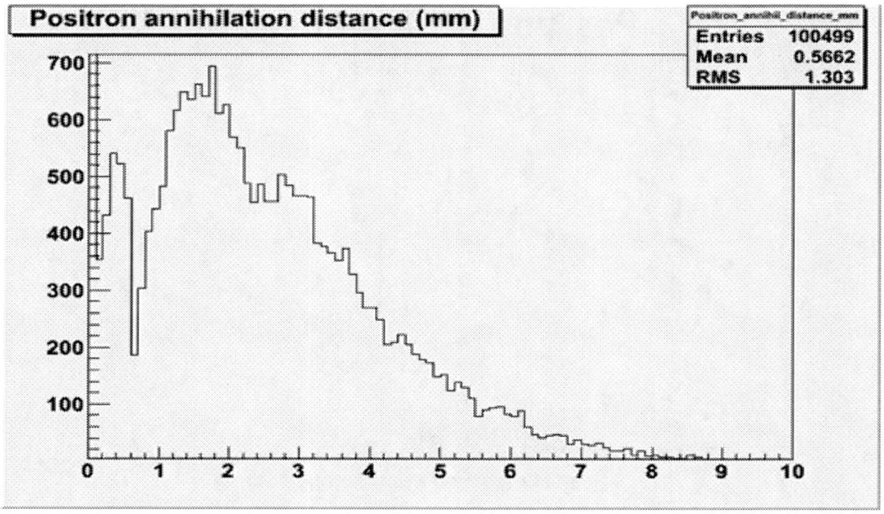

Figure 9. Simulated positrondistance in water cylinder from ^{124}I ∅1 mmline source. Initial high peak and unstable values until 0.5 mm, are due to unstable simulation results in the space of the source.

4. Biologist World

4.1. Uniformity of Calibration, Why Does the Same FDG Rat Brain Scan Appears So Much Different on Each Scanner?

I was always puzzled why the same control animal (rat or mouse) [^{18}F]FDG-PET scan done on different microPET scanners look so different. An example structure to look at is coronal view of hippocampal formation in the brain. On some scanners, one would see an enhanced (i.e. with higher FDG uptake) striatal area, on some other scanners, the cortical area is more pronounced. Let us examine some possible differences of this effect.

We must remember that PET image is an image reconstructed from individual events of positron annihilation. On the way to reconstruction, there are many steps, from a relatively well detected annihilation event to a highly processed image. An individual annihilation event is detected with varying probability at different positions in the field of view, and this probability falls rapidly from the center of the field of view to its axial edges. This effect is corrected in the reconstruction by proper calibration of the system. To check the uniformity of such calibration, a scan of a cylinder with water diluted tracer is done. By definition, Standard Uptake Values (SUV) must be one for such scan. SUV is defined as detected concentration of activity divided by the volume of the phantom (patient) and divided by injected activity. An example of such a scan for mouse cylindrical phantom is shown in Figure 10.

SUV of this phantom is 0.96±0.06 (standard deviation), so within error corresponds to expected SUV =1. The VOI diameter was 12 mm, to stay away from the blurred edges resulting from finite resolution of the system. The edge showed in Figure 10 a rises from 10% to 90% of the maximum in 4.3 mm. The result shows that non-uniformity of this scanner is about ±6.3%. As some biological effects, which are measured in PET give only about 20% difference between the studied and control group – than the non-uniformity of the scanner calibration may substantially blur the results.

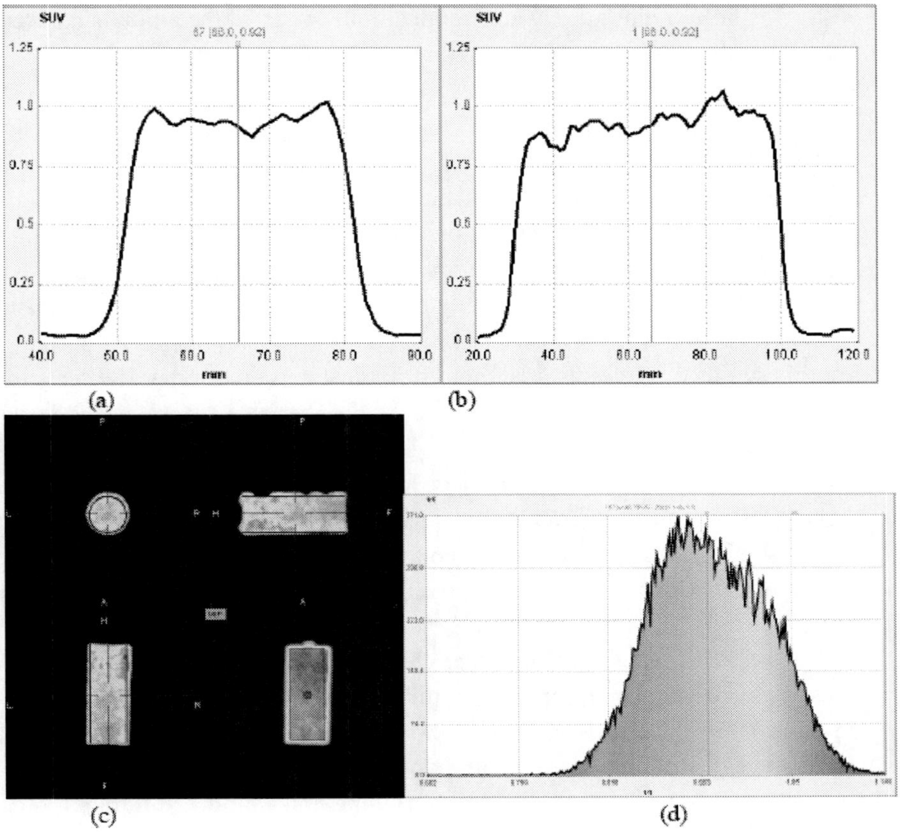

Figure 10. 83 MBq, 1 hour, [^{18}F]FDG-PET scan of a 49 ml cylindrical (∅30x70 mm) water-filled phantom in plexiglass container. Cylinder axis along axis of the Mosaic microPET scanner.Cylinder located in the center of the field of view. (a)Horizontal and (b) longitudinal reconstructed activity profile with the cursor at the center of the field of view. (c) Image of the cylinder in three planes and in Maximum Intensity Projection (MIP). (d) Histogram of SUV values within volume of interest (VOI) shown in blue in (c) cylinder with color coding corresponding to the image (c). By definition, the value of the reconstrucion should be 1 SUV.

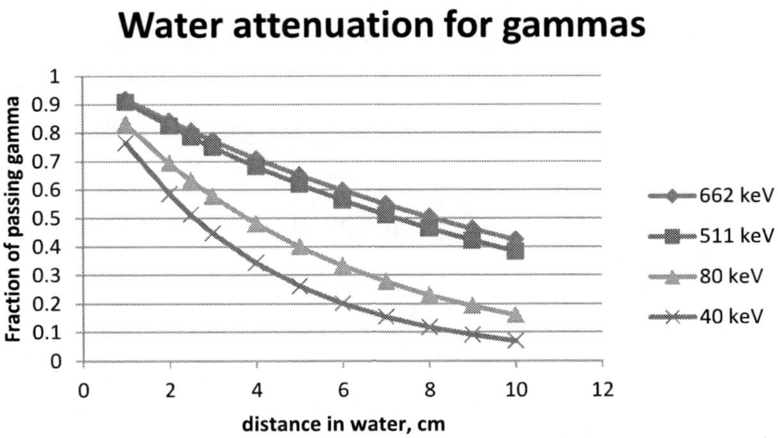

Figure 11.Water attenuation for several γ and X-ray energies at a distance corresponding to possible lines of response in the small animal.

Some of the 511 keV annihilation γ are stopped in the body of the animal. To account for that effect, transmission scan is sometimes acquired, especially for rat, because for the mouse, the effect is smaller. Transmission scan is now usually done with medium energy X-rays (80 keV) and used in reconstruction of the image, assuming that attenuation coefficient propagates as water one. Older systems used ^{137}Cs source of 662 keV, which is more appropriate, but more difficult to handle. In humans, usually a line source of 511 keV is rotated around the patient to measure attenuation coefficient.

In 3D reconstruction of the image, there are some oblique lines of response which pass larger distances in the animal body than mentioned in Section 2.1.. The effect is illustrated in Figure 11.

Apart from these physical effects, many physiological and other type of effects influence the content of the image:

1. Animal temperature during the scan. For long time scans, the bed of the animal must be heated, as without it, the animal is cooled to a non-physiological temperature. In [^{18}F]FDG-PET scans, one can easily see the effect by appearance in the image of the brown fatin the neck of the animal, which starts to consume glucose, and hence, the tracer.
2. In [^{18}F]FDG-PET scans, an important factor is the fasting of the animal. Fasting increases glucose consumption, but is quite difficult to control and introduces inter-animal error by itself. For some other tracers, blocking of the receptors by intraperitoneal prior to the scan is introduced to enhance tracer uptake.
3. Anesthesia is now usually done by isoflurane inhalation. Isoflurane is a potent GABA receptor agonist so brain studies may be affected by it. Glucose consumption in the brain in [^{18}F]FDG-PET scans is reduced by about three times during anesthesia.
4. If animals are exposed to fear during preparation to microPET scan, it may also greatly influence brain activity and hence, the [^{18}F]FDG-PET brain image.
5. We do microPET scans of animals during their normal time of rest, which corresponds to our night, and minimum activity period. For behavioral animal studies which are then examined in PET, it is a factor of importance.

As acquiring images is usually difficult, the usual number of animals per group is four, which is quite a low number for statistical analysis. If possible, the number of the animals should be about six to eight.

4.2. I Have a New Functional Property to Image – How Do I Select or Invent a Tracer for It?

When a functional property up to the moment assayed *in vitro* is to be imaged *in vivo*, many problems arise. Here we show two examples of possible approaches.

Imagine that you want to image NMDA receptors *in vivo*. What is quite easily achievable in vitro seems to be a very difficult task *in vivo.* Such a tracer needs to be specific for NMDA receptor, pass through blood brain barrier and not to be too fast metabolized, to allow imaging.

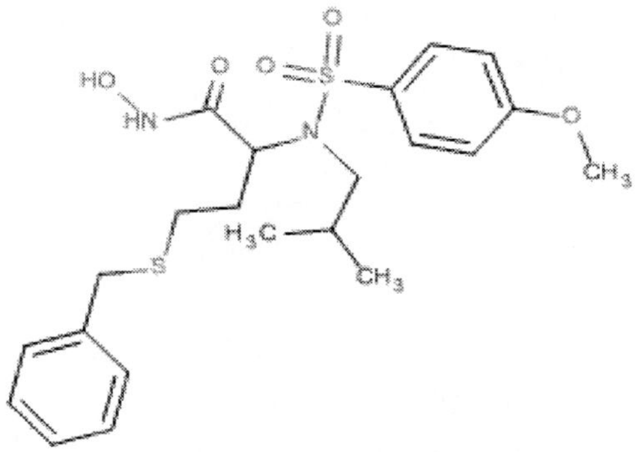

Figure 12. 4-benzylsulfanyl-N-hydroxy-2-[isobutyl-(4-methoxybenzenesulfonyl)-amino]-butyramide potent MMP-9 selective inhibitor. The part responsible for selective binding to MMP-9 is –S-CH_2-C_6H_5 at lower left corner [23].

Neither NMDA compound nor glutamate are viable tracers for NMDA receptor, as they are released from the receptor upon conformational change when the neurotransmitter binds.

Natural choices for tracers are antagonists of such receptors. Electrophysiology offers a wide range of various subtypes of NMDA receptor antagonists at various sites of the receptor (ifenprodil, MK-801)[18]; however, none of them or their derivatives were successful as a tracer when labeled with ^{11}C or ^{18}F (example - [19]). The tracer is injected normally at doses much lower than therapeutically active, so adverse effects would be minimal. Most of the reported compounds failed to be a good PET tracer.

Matrix metalloproteinases (MMP), a group of more than 20 endopeptidases, especially MMP-9 was shown to be highly involved in cancer cell proliferation [20]. Then it was shown to have role in memory and learning [21]. MMPs share active site architecture. Specificity of the protein is achieved by subunits other than the one containing active site of the enzyme. With that, there are quite a number of broad spectrum MMP inhibitors, which interfere with active sites, but are not specific for MMP-9 [22] except for S24994 (4-benzylsulfanyl-N-hydroxy-2-[isobutyl-(4-methoxybenzene-sulfonyl)-amino]-butyramide compound 6a in [23] shown in Figure 12).

This compound was never tried as a PET tracer, but potential sites for ^{11}C incorporation are three methyl groups. OH group rather cannot be replaced by ^{18}F, as the oxygen of this group binds Zn^{2+} ion and hydrogen binds to Glu residue of the MMP peptide.

References

[1] Balcerzyk, M., et al., *Measurement Science and Technology* 20, 104011 (2009).

[2] Bao, Q., Newport, D., Chen, M., Stoutand, D.B. Chatziioannou, A.F. *J. Nucl. Med.* 50, 401 (2009).

[3] Lerche, C. *Universidad de Valencia* (2006).

[4] Phelps, M.E., Ed., *PET: Molecular Imaging and Its Biological Applications,* Springer London, London, (2004).

[5] Judenhofer,M.S., et al., *Nat. Med.* 14, 459 (2008).

[6] Bruno,R., et al., *Journal of endocrinological investigation* 27, 1010 (2004).

[7] Tazebay, U.H., et al., *Nat. Med.* 6, 871 (2000).

[8] Jauregui-Osoro, M., et al., *Eur. J. Nucl. Med. Mol. Imaging* 37, 2108 (2010).

[9] PMOD - Biomedical Image Quantification, 3.206, *PMOD Technologies,* www.pmod.com.

[10] Schambach, S.J., Bag, S., Groden, C., Schillingand, L., Brockmann, M.A. *Methods* 50, 26 (2010).

[11] Grignon, C., et al., *Nucl. Instr. Meth. A* 571, 142 (2007).

[12] Cho, Z.H., et al., *Journal of Nuclear Medicine* 16, 1174 (1975).

[13] Levinand, C.S., Hoffman, E.J. *Phys. Med. Biol.* 44, 781 (1999).

[14] Craigand, S.L., Edward, J.H. *Phys. Med. Biol.* 45, 559 (2000).

[15] Derenzo, S.E. *IEEE Trans. Nucl. Sci.* 33, 565 (1986).

[16] Shibuya, K. et al., *Phys. Med. Biol.*52, 5249 (2007).

[17] Jan, S., et al., *Phys. Med. Biol.* 49, 4543 (2004).

[18] Sobrio, F., Gilbert, G., Perrio, C., Barréand, L., Debruyne, D. *Mini-Reviews in Medicinal Chemistry* 10, 870 (2010).

[19] Labas, R., et al., *European Journal of Medicinal Chemistry* 46, 2295 (2011).

[20] Bergers, G., et al., *Nature Cell Biology* 2, 737 (2000).

[21] Nagy, V., et al., *J. Neurosci.*26, 1923 (2006).

[22] Jourquin, J., et al., *European Journal of Neuroscience* 18, 1507 (2003).

[23] Hanessian, S., Moitessier, N., Gauchetand, C., Viau, M. *J. Med. Chem.* 44, 3066 (2001).

Index

C

D

E

F

G

H

I

K

L

O

P

Q

R

S

T

U

V

W

X